THE Kitchen SCIENCE COOKBOOK

The Kitchen SCIENCE Cookbook

By Dr Michelle Dickinson

Executive Producer: Joe Davis
Design: Suburban Creative Limited
Photography: Magic Rabbit Limited

PARTICULAR BOOKS
an imprint of
PENGUIN BOOKS

PARTICULAR BOOKS

UK | USA | Canada | Ireland | Australia
India | New Zealand | South Africa

Penguin Books is part of the Penguin Random House group of companies
whose addresses can be found at global.penguinrandomhouse.com.

First published in New Zealand by Nanogirl Labs Ltd 2018
First published in Great Britain by Particular Books 2019
002

Copyright © Nanogirl Labs Limited, Auckland, New Zealand, 2018

The moral rights of the authors has been asserted

Printed and bound in Italy by Printer Trento S.r.l

A CIP catalogue record for this book is available from the British Library

ISBN: 978-0-241-39558-5

www.greenpenguin.co.uk

Penguin Random House is committed to a
sustainable future for our business, our readers
and our planet. This book is made from Forest
Stewardship Council® certified paper.

Dedicated to
the future
problem solvers
of the world

~ stay curious!

CONTENTS

7

CONTENTS Continued....

As an engineer and a science communicator I often meet people who tell me that they wish they were 'good at science too'. Their understanding of 'science' often relates back to their school experience - they see it as a subject they used to study, rather than as a way of life.

One particular lady, a mother who brought her children along to one of our live science shows, said exactly this to me - and at the same time handed me a cake she had baked for our team. She was a scientist. She could follow a recipe and adjust the way she cooked to affect the outcome or solve a problem - she just didn't think about it in that way. That conversation was the inspiration for this book, and the start of three years' research.

Using the scientific knowledge I've gathered from my experience in research laboratories, I've spent three years in the kitchen at home, trying different experiments. After lots of trial and error, we have compiled the top 50 of these experiments in this book, each presented as an easy-to-follow recipe, just as you might find in a cookery book.

The experiments in this book are easy to do. You need no previous scientific experience, and they only require ingredients and equipment that are commonly found in the home.

All the recipes in this book are designed to be enjoyed by the whole family. They explore the worlds of physics, chemistry and biology while also using the construction and building skills vital for budding engineers.

Scientists and engineers are the problem solvers of our world. This book is written to help develop these skills in a way that is fun - and sometimes even edible! It has also been created with a mission in mind. We believe that science should be for everyone, and to that end we are donating many thousands of these books to families, schools and charities around the world that would not otherwise have the opportunity to explore science in this way.

You don't need any qualifications to be a scientist - just curiosity, a willingness to try things, and to get things wrong from time to time... that's often where you'll learn the most!

I hope that this book will inspire you to explore science further. We have loved creating and testing these recipes for you!

Michelle

- Dr Michelle Dickinson

EACH OF THE 50 recipes IN THIS BOOK IS *a science experiment* WHICH YOU CAN COMPLETE AT HOME USING ONLY *common ingredients,* MOST OF WHICH YOU WILL PROBABLY ALREADY HAVE IN YOUR *kitchen.*

Each recipe contains a simple introduction explaining what the experiment is about, as well as icons identifying important details - such as if the recipe is edible, if it requires time to 'set', or if there are any safety issues that you need to be aware of.

The equipment and ingredients lists specify all the materials you need to complete the recipe. The instructions list provides a step by step guide to all the procedures.

Before starting the recipe, ask yourself what you think may happen - and why. Scientists call this a hypothesis and it's a great way to see if you can predict a result before you actually carry out the experiment. It may help to have a notebook to write down your hypothesis, and to record any observations you make along the way.

The process of science is just as exciting as the results! Make a point of stopping at each step, and observe if anything is changing before you proceed with the next step. Sometimes the colour, temperature, texture, or smell may change as you add different ingredients.

'The Science Behind' section of each experiment is to help explain the results from the recipe. However, do remember that different brands of ingredients may differ slightly, so some recipes may need a bit of tweaking or further experimentation.

After you have completed a recipe, go back to your hypothesis and see if your results were the same as you expected.

Don't worry if you get the recipe wrong, or if it doesn't turn out quite how you planned – this is what being a scientist is all about! Mistakes and troubleshooting are the key ingredients for problem solving. Some of the world's best inventions have been discovered from things going wrong in the lab!

The 'Explore Further' section is designed to help you to think more about how the recipe works. It suggests extra tasks and challenges to help you see how changing the variables can change the results.

In general, we have tried to organise the recipes in each chapter to start with the easiest to do, becoming progressively more challenging through the section.

A series of icons are used throughout the book to give you helpful guidance and important safety information.

Basic KITCHEN SCIENCE Rules

To keep safe and enjoy all the recipes in this book, it's good to follow some basic kitchen cooking rules.

1. Wash your hands with soap and warm water before touching any food or ingredients. Many of these recipes are edible, so it's important that your hands are clean - this way, everybody can safely enjoy the tasty results.

2. Wear an apron and ensure that any long sleeves are rolled up. If you have long hair, tie it back to reduce the risk of it catching in equipment or any loose hairs falling into the food.

3. Before starting a recipe, read the instructions from beginning to end to make sure that you have all the ingredients and equipment to hand.

4. To ensure a greater chance of a successful experiment, take time to measure every ingredient carefully. Refer to the 'Measuring and Conversions' section in this book.

5. A timer is a useful tool, both in the kitchen and the laboratory. Many modern phones have a timer function.

6. When touching hot saucepans or baking trays, always take care and use heat resistant gloves to avoid burning your hands.

7. When you have finished your experiment, do not forget to clean up and put all your equipment away. A tidy lab is a safe lab!

Heat and flames

Some of the recipes in this book require heating in an oven, on a stove or using an open flame.

We advise parents and caregivers to ensure that young people in their care are familiar with the safety rules below prior to commencing any experiment involving heat.

Adult supervision is important for all these experiments.

Ensure that long or baggy clothing is not worn when carrying out these experiments, and that long hair is tied back.

Oven Safety

Always use heat resistant oven gloves and take care when handling hot baking trays.

Stand to the side when opening the door of a hot oven to prevent the initial blast of hot air burning your face.

Stovetop

Make sure that all pan handles are turned to the side to minimise the risk of pans being knocked off the stove.

Never leave a pan unattended or filled too high on a stovetop as it can quickly boil over.

Take care when stirring boiling liquids as the rising steam can burn. Use a long handled spoon to minimise this risk.

Use heat resistant gloves to hold the handle of a hot pan and when finished, rinse the pan off in cold water to reduce the risk of others picking it up while still hot.

Microwave

Make sure that only microwave-safe plastic or glass dishes are used in the microwave. Never use metal foil to cover food in the microwave.

When removing a lid or cover from a liquid that has been heated in the microwave, take care with escaping steam as it can easily burn.

Use heat resistant gloves when removing items from the microwave - bowls and plates can be very hot at the bottom, even if they feel cool at the top.

Keep an eye on liquids being heated in the microwave as they can easily boil over if left unattended.

Open Flames

Ensure that all flammable materials have been removed from the area and that there are no draughts in the room before lighting a candle or gas element.

Never leave an open flame unattended and always have an adult present when lighting a candle, either using long matches or a long lighter.

Make sure the candle has a sturdy base to prevent it falling over and keep a bowl of cold water close by to put out a rogue flame if needed.

Knives and sharp things

Safety is very important in any kitchen or laboratory. Some of the recipes in this book use scissors, knives, graters or a blender, all of which require care, especially when used by children. Following is some detailed safety advice for these implements which, if you are a parent or caregiver, you may already be familiar with. We strongly recommend, however, that all young scientists are taught the safety information below before commencing any of the experiments.

Knives

Ensure hands are dry and clean so they do not slip easily and that children are supervised at all times when using sharp tools. When teaching a child to hold a knife, encourage the use of a 'pinch grip' - where the thumb and forefinger hold firmly onto the sides of the blade at the end of the handle. This helps to ensure control. From there, move on to the 'pointer grip' - where the index finger is placed along the top of the knife to help steady the hand.

Ensure that objects being cut with a knife are held firmly, with the fingers out of the way of the blade. A claw grip can be used where the fingers are tucked underneath the hand to minimise the chance of cutting the fingertips.

Scissors

To teach a child to use scissors, begin by demonstrating how to hold scissors correctly - with the thumb in the upper hole, and the index and middle fingers in the lower hole. Teach the child how to cut away from the body, using a forwards motion with each consecutive snip. Children will develop fine motor skills at different rates, so it is important to teach each child the skills that match their development and help them with more complicated cutting tasks when necessary. Ensure the child is familiar with the correct way to walk with scissors - by holding the blades together in a fist with the handles sticking out on top.

Grater

A grater can be just as sharp as a knife, and should be used with a slow and careful grating motion. Ensure that the object being grated is large enough and that the child stops grating well before they get to the end of the ingredient. Teach the child to push the object to be grated down against the blade, ensuring their fingers and knuckles are tucked away to prevent catching them on the grater.

Blender

Blenders contain sharp blades so care should be taken when initially fitting the blades.

Ensure that hands are dry before using a blender. Explain that the lid is always kept on the blender when in use and that a spoon or spatula must never be placed inside the blender without it being switched off.

Measuring & CONVERSIONS

Abbreviations

tsp   = teaspoon

Tbsp = tablespoon

cm   = centimetres

m   = metres

in   = inches

ft   = feet

g   = grams

ml   = millilitres

Following a recipe and carrying out a science experiment each require careful measurement of all ingredients and chemicals.

In cooking, different units of measurement - such as imperial or metric - are used in different countries.

In science, a standardised system of measurement known as the International System of Units (commonly abbreviated to SI units) is used all over the world.

Measuring

When measuring large volumes of liquids use a measuring jug, keeping your eye at the same level as the liquid in the jug.

For smaller liquid volumes use a measuring spoon.

When measuring dry ingredients use a weighing scale, ensuring that you zero the scale with your empty measuring container already in position on the scale.

For smaller quantities of dry ingredients, use a measuring cup or spoon. Ensure that you fill the cup or spoon with the ingredient, then level with the back of a knife or spatula.

In this book we will provide approximate conversions for common measurements of volume and weight. For you reference, here is an approximation conversion chart:

Liquids

5 millilitres = 1 teaspoon

15 millilitres = 3 teaspoons = ½ fluid ounce

60 millilitres = 4 tablespoons = ¼ cup

120 millilitres = 4 ounces = ½ cup

250 millilitres = 8 fluid ounces = 1 cup

480 millilitres = 16 fluid ounces = 1 pint

1 litre = 2 pints = 1 quart

4 litres = 128 fluid ounces = 1 gallon

Solid Particle Ingredients (sugar, flour)

4 grams = 1 teaspoon

12.5 grams = 3 teaspoons = 1 tablespoon

200 grams = 1 cup

Temperature

140°C = 275°F = Gas Mark 1

150°C = 300°F = Gas Mark 2

170°C = 325°F = Gas Mark 3

180°C = 350°F = Gas Mark 4

190°C = 375°F = Gas Mark 5

200°C = 400°F = Gas Mark 6

220°C = 425°F = Gas Mark 7

230°C = 450°F = Gas Mark 8

240°C = 475°F = Gas Mark 9

Keeping NOTES

Scientists usually keep a log of their experiments to record their studies and results.

The scientific method typically involves the scientist asking a question, then writing down observations while performing the experiment.

Getting a notebook to keep track of your experiments is a great way to record observations and also to see how changing parts of your experiments can change your results.

Example questions you could answer in your science notebook include:

What day and time did you start the experiment?

This helps with experiments that can take a long time to grow, such as 'Rock Candy', or that take time to set, such as the 'Erupting Volcano' or 'Edible Earthworms' experiments.

What questions do you need to ask about the experiment?

For example, you could ask: 'What will happen if I add vinegar to milk?' or 'How long will it take for the food colouring to spread out?'

What do you think will happen in the experiment?

This is what scientists call building a hypothesis or an explanation for what happens in a science experiment.
It is usually a guess about what might happen, based on what you know about the ingredients.
For example, if you know that milk curdles, you might guess that the vinegar will curdle the milk.

What happened in your experiment?

Write down or sketch the results of your experiment to keep a record of what you observed.

Were the results the same as you thought they would be?

Did everything go to plan? If not, why not?
Is there something you could change to get a different result?

Did the experiment fit your hypothesis?

How close to your hypothesis were the results?
Did the results align with what you thought would happen in the experiment?

What is your Conclusion?

Why do you think what happened in your experiment happened?

What would you do differently next time?

Are there different things that you could do in your experiment to test your hypothesis?

By keeping a record of your recipe experiments you can refer back to the results when you try out other experiments. Sometimes you will see that the science is connected.

Allergies & Sensitivities

Whether due to personal choice, sensitivity or allergy, some of the recipes in this book may not be suitable for all.

Gelatin is used as a crosslinking ingredient in several of our recipes. Gelatin is sourced from animals and therefore may not be suitable for vegetarians. There are several alternatives that can be used as a substitute. Depending on your needs, kosher gelatin or vegetarian agar powder can be used in the same amounts as standard gelatin in all recipes. Carrageenan or Irish moss can also be used by substituting 28 grams of carrageenan for every teaspoon of gelatin.

White flour which contains gluten is used in the 'Bread in a Bag' recipe. All the other edible recipes use cornflour/cornstarch, which is gluten free.

Nuts are used in the 'Milking Nuts' and 'Confectionary Candle' recipe.

Dairy products including milk and cream are used in the 'Milk Sculptures', 'Instant Ice Cream', 'Making Butter', 'Scrumptious Slime', 'Microwave Cheese' and 'Edible Earthworms'. The cream can be excluded from the 'Edible Earthworms' experiment without changing anything other than the opacity of the finished product.

The only edible recipe that uses eggs is 'Fun Foamsicles'.

LOOK
for these icons

Fire - This recipe requires a heat source to complete - either an open flame, a microwave oven, a stovetop or a conventional oven. Adult supervision is needed when working with heat sources, and appropriate care must be taken such as wearing heat resistant gloves, to minimise the risk of burns. We recommend that long hair be tied back when working with heat sources, particularly with open flames.

Sharps - This recipe requires the use of sharp tools, including knives and scissors. Adult supervision is needed when working with sharps. Please read the advice in the 'sharps safety' section of this book carefully before attempting these recipes.

Edible - The end product of this recipe is edible! If you wish to eat it, please take all due care for food safety during preparation as you would in a normal kitchen. Be sure to check the ingredients carefully if you have any allergies or food sensitivities.

Outside - This recipe may require sunlight, a large area, or could be messy. We recommend that this recipe is carried out in a safe outdoor space.

Time Needed - These recipes need a bit more time than the others to finish, often to allow ingredients to set, or to dry before they are ready for use.

Section 1

COLOURFUL

EXPERIMENTS

COLOURED Flowers

Add any colour you like to decorate your flowers with this beautiful experiment that shows you how quickly plants can drink.

Equipment & Ingredients

- Scissors
- Jars or glasses
- White flowers
- Food colouring - several different colours
- Water

Instructions

1. Fill each jar or glass 3/4 full with water.
2. Add 10 to 20 drops of food colouring to each jar - a different colour for each jar.
3. Using scissors, carefully trim the stem of the flowers at a 45 degree angle to maximise the cut stem surface area.
4. Place each flower in a jar, checking every few hours to see if the flowers petals change colour.

The Science Behind Coloured Flowers

Plants draw water up through their stems through a process known as transpiration. As the water moves through the plant, it eventually evaporates from the leaves and flowers through openings called stomata. As the water evaporates, a pressure difference is created which helps to pull more water into the plant. In a similar way, sucking through a drinking straw creates a pressure difference which pulls liquid up through the straw. Because flowers are usually only put in water, which is colourless, we do not see the fluid passing through. However, adding colour to the water allows the path of the water to be tracked as it passes through the plant.

The flowers can be left in the coloured water for several days and observed as different amounts of colour appear on the petals. The darker the colour, the more transpiration there has been at that point.

Explore Further

» At the end of your experiment, cut the stem lengthways down the middle and examine to see if there is any food colouring inside the stem.

» Try testing what affects the transpiration rate of the flowers by placing different jars in areas with different conditions - such as a sunny window ledge, a dark cupboard or a humid bathroom - and seeing which flower ends up with the most colour.

» What does this tell you about the evaporation rate?

Glass RAINBOWS

All the colours of the rainbow are contained within sunlight. Explore the refraction of light using a simple glass of water.

Equipment & Ingredients

- Water
- A transparent colourless glass
- White paper
- A sunny day

Instructions

1. Fill the glass 3/4 full with water.
2. Lay the sheet of paper on a flat surface close to a sunny window.
3. Place the glass of water on the paper and slowly lift it up, away from the surface of the paper.
4. Look carefully at the light passing through the water onto the paper - can you see any colours of the rainbow?
5. Try increasing the height at which the glass of water is held above the paper - can you see more colours or not as many?

The Science Behind the Glass Rainbow

In the sky, rainbows form as arcs of colour on days where there is both rain and sunshine. Rainbows can form in lots of other places too. Light from the sun is made up of white light. White light is light which contains all colours in roughly equal quantities.

Light travels more slowly through water than through air. As the sunlight shines down from the sky, it moves quickly through the air and then passes through raindrops where the water slows it down. Each colour within the white light has a different wavelength and bends or refracts as it passes through the water. Because each wavelength bends by a slightly different amount, each separate colour can now be seen.

When the light is shone through the glass of water it also refracts or bends, and the separated colours can be seen on the piece of paper. Depending on the strength of the sunlight, and how focused it is, all or some of the red, orange, yellow, green, blue and violet colours that make up light may be seen.

Explore Further

» Can you make a rainbow with a torch or flashlight instead of the sun?
» What happens if you place a mirror at an angle inside the glass to catch more of the sunlight?
» Can you find rainbows anywhere else around the house?
» If you cover the window with paper, and only let a thin slit of light pass through your glass, does it change the strength of your rainbow?

Wicking WATER

The power of wicking (absorbing or drawing off liquid) can be used to move coloured water from one glass to another, creating a beautiful ever-changing piece of art.

Equipment & Ingredients

- 6 small, wide-mouthed jars or glasses
- 6 paper towels
- Scissors
- Food colouring - 3 different colours

Instructions

1. Fold each sheet of paper towel twice along its length forming a narrow strip.
2. Arrange the glasses into a circle and place the end of one paper towel into the base of one glass.
3. Fold the paper towel over so that it sits over the mouth of the glass and touches the base inside the glass adjacent to it. You may need to trim the end of the paper towel depending on the height of your glasses.
4. Repeat using the other paper towels until each glass is connected to the next with a paper towel.
5. Add several drops of food colouring to each alternate glass and fill with water, leaving the remaining glasses empty.
6. Watch as the coloured water wicks across the paper towel and into the empty glass next to it.
7. In some of the empty glasses the two colours should eventually mix to create new colours.
8. After the experiment, unfold the paper towels and leave to dry outside in the sun for a beautiful piece of science art.

The Science Behind Wicking Water

Paper towels are designed to soak up fluid, which is why they are used to clean up spills in the kitchen. The water travels across the paper towel using a process called capillary action. Capillary action is the ability of a liquid to flow upwards through small spaces, against the force of gravity. Paper towels are actually made from the cellulose fibres found in plants. In this experiment, the water flows upwards through the tiny gaps between the cellulose fibers which act as capillary tubes, pulling the water upwards over the mouth of the glass and into the next glass. The food colouring helps us to see where the water is travelling from and to. You can see which fluid wicks faster by observing how much - or little - of each colour mixes in the middle.

Explore Further

» What happens if you roll the paper towels into a tube shape instead of folding them? Why do you think this is?
» Does the experiment change if you use warm water instead of cold water?
» Do tall glasses take a longer or shorter time to wick the water than shorter glasses? What is it about the height of a glass that might make a difference?

The coloured water travels
across the paper towels
using a process called
capillary action.

Milk SCULPTURES

This experiment takes two liquids and transforms them into a plastic that can be moulded into any shape.

Equipment & Ingredients

- Sieve
- Spoon
- Paper towels
- Bowl
- 1 cup of milk
- 15 ml (1 Tbsp) white vinegar

Instructions

1. Heat the milk on the stove or in the microwave until it is warm but not boiling.
2. Add the vinegar to the milk and stir vigorously with a spoon for one minute - you should see lumps starting to form.
3. Pour the milk mixture through the sieve to drain off any liquid.
4. Leave the strained lumps in the sieve until they are cool enough to touch.
5. Spoon the lumps onto a paper towel and squash together to blot off any remaining fluid.
6. Mould the lumps into any shape that you like, then leave in a warm, dry place to harden into a solid milk sculpture.

The Science Behind Milk Sculptures

Milk is made up of a protein called casein. When an acid such as vinegar is added, it changes the pH or acidity of the milk. This change causes the casein molecules to unfold from their balled up structure to a long chain structure. In the food world, this is also known as curdling milk. The long casein molecules link together which results in a material called a polymer or a plastic. Polymers are flexible solids that can easily be shaped and, when left to dry, become less flexible and harden into a set shape.

The same process was used in the 1900s to make ornaments, and was even used to make jewellery for Queen Mary of England. Today casein polymers are used to make artists' paints and some glues.

Explore Further

» Does the experiment work with other dairy products like cream or yoghurt?
» Can you use other acids - such as lemon juice - instead of vinegar, to create the same effect?
» Instead of making a sculpture, you can roll the solid milk onto a plate and make an impression of your hand.

Window WOBBLERS

Stained glass can transform transparent windows into art using colourful glass shapes. In this experiment you can create your own colourful window art using the science of crosslinking.

Equipment & Ingredients

- Drinking straw
- Shallow tray or dish
- Cookie cutter
- Toothpick
- Spatula

- Paper towel
- 15g (1 1/2 Tbsp) gelatin
- 400ml (1 3/4 cups) boiling water
- Food colouring

Instructions

1. Add the gelatin to the boiling water and stir well until dissolved.
2. Pour the water into the tray and allow to set in the fridge for 4 hours.
3. Once set, use a straw to make several holes spaced across the gelatin. Use a toothpick, if necessary, to pull out the gelatin plug left by the straw.
4. Fill each newly created hole with a drop of food colouring. Let the gelatin sit for 4 more hours in the fridge, checking often to watch the food colouring spread.
5. Use cookie cutters to cut shapes out of the gelatin sheet. Carefully remove the shapes from the tray using a spatula.
6. Dab the shapes dry with a paper towel to remove any residual food colouring, then place them on a window, holding each one for a few seconds until it sticks.
7. Watch the shapes change as the water evaporates over the next few days.
8. When they are dry, peel the shapes off the window and rehydrate by soaking in water for a few hours.

The Science Behind the Window Wobblers

Window wobblers use gelatin to transform the liquid water into a solid gel. This gel can still absorb fluid and you may have noticed the drop of food colouring spread out over time from the hole that was created, moving further into the jelly. This is due to a process called diffusion, where molecules move from areas of high concentration to areas of low concentration. The high concentration of food colouring spreads out to areas of the gelatin mixture that do not have as much.

Since the window wobbler is mostly made of water, the heat from the sun as the wobbler sits on the window causes the water to evaporate, making it thinner and less flexible. By dropping the jelly back into water, water can be re-absorbed, enabling the jelly to swell back up to a wobbly state. The wobbler can then, of course, be stuck onto the window again.

Explore Further

» Can you measure the rate of diffusion of food colouring as it moves away from the hole created by the straw? Does the rate speed up or slow down over time?

» Will the amount of sunshine a wobbler is exposed to change how quickly the water evaporates from it? Can you measure this difference?

» Does the wobbler shape rehydrate better with warm or cold water? Why do you think this is?

Density DISCS

Amaze your friends as you balance fluid solutions of different densities on top of each other to create a beautiful, coloured stack of science!

Equipment & Ingredients

- Drinking straw
- 4 tall, colourless glasses
- Tablespoon
- Hot water
- Food colouring - 4 different colours if available
- Sugar

Instructions

1. Place the 4 glasses in a line. Leave the first glass empty. Add 15g (1 Tbsp) sugar to the second glass, 30g (2 Tbsp) of sugar to the third glass and 60g (4 Tbsp) of sugar to the fourth glass.

2. Add 3 drops of food colouring to each glass, a different colour for each glass if possible. If you do not have 4 different colours, you can mix colours together to form a different colour.

3. Pour 60ml (4 Tbsp) of hot water into each glass. Stir each solution until all the sugar has dissolved.

4. Place a straw all the way to the bottom of the third glass. Hold your finger over the top to keep the liquid in the straw, then move it over to the fourth glass.

5. Hold a tablespoon upside down inside the fourth glass, placing it so that the tip of the spoon is against the inside edge of the glass, slightly above the first layer of liquid.

6. Using the fluid filled straw as a pipette, release your finger from the top of the straw, alllowing the liquid from the third glass to drop over the back of the spoon.

7. Keep transferring the liquid from the third glass to the fourth glass until you can see a new layer of colour sitting on top of the fluid in the fourth glass.

8. Repeat with the fluid from the second glass, and then the first glass, until you have all the fluid layers stacked on top of each other.

The Science Behind Density Discs

Density is a measure of how much mass there is in a given volume. Each of the glasses contains the same volume of water but a different amount of sugar. When the sugar molecules dissolve in the water, they increase the mass in the water which increases the density. The more sugar in the water, the more dense the mixture or solution. Solutions with less density float above solutions with higher density, so by stacking the mixtures in order of density you can get them to float on top of each other. If the fluids come together with too much force they can mix, so the spoon is used to reduce the force of the liquid as it is poured into the glass. The straw ensures that only small volumes are transferred at a time, which reduces the fluid interaction force and increases the stacking success.

Explore Further

» Can you use other materials that dissolve in water - such as salt - to create different density discs?

» What happens if you use cold water instead of hot water? Will the experiment still work?

» What will happen if you stir the density disc column? Will the discs settle back into their stacked states? Why do you think that is?

Liquids of different density can be made to float on top of one another with beautiful results.

Section 2

CONSTRUCTION

EXPERIMENTS

Marshmallow CATAPULTS

Catapults have been used throughout history to launch objects at and over the walls of castles by converting potential energy to kinetic energy. This experiment uses the same theory - but with objects you can eat!

Equipment & Ingredients

- 2 elastic/rubber bands
- 1 plastic teaspoon
- Masking tape
- 7 wooden skewers
- 5 large marshmallows

Instructions

1. Create a triangular base with three of the wooden skewers, joining them with a marshmallow at each corner.
2. Place one skewer upright into the centre of each of the three marshmallows used in the base, then bring the tips of the skewers together to transform your triangle into a pyramid.
3. Secure the three skewer tips with an elastic/rubber band, then place a marshmallow over the top.
4. Use masking tape to attach the spoon to the end of the final skewer.
5. Feed the spoon-skewer combo through the centre of the pyramid, placing the wooden end into the base triangle marshmallow at the front of the structure and holding the spoon end at the back of the structure.
6. Place the second rubber band over the top of the pyramid and loop it under the spoon to hold the spoon off the ground or table.
7. Place a marshmallow on the end of the spoon, pull it backwards against the rubber band and let go!

The Science Behind the Marshmallow Catapult

Catapults work by converting energy from one type to another and transferring this energy from one object to another. When the spoon is pulled back against the rubber band, energy is added to the catapult system. This energy is stored as potential energy in both the spoon and the rubber band. The further the spoon is pulled back, the more potential energy is stored. When the spoon is released, the potential energy converts to kinetic energy (energy in motion) and the spoon propels forward, releasing the energy stored in it. This energy is transferred from the spoon to the projectile marshmallow sitting in the spoon, and causes the marshmallow to fly through the air. Because the spoon and skewer combo are long and bendy, they act as a lever, pivoting on the marshmallow base and enabling the projectile to be propelled a long way relative to the small amount of effort.

Explore Further

» What happens if you shorten the length of the spoon-skewer lever? Why do you think that is?
» Can you construct a catapult using a shape other than a pyramid? How about a square or a rectangle?
» How does changing the length of the rubber band affect how the marshmallow flies?
» Can you aim the marshmallow so that it lands in a bowl across the table?

Fish in THE TANK

This experiment demonstrates how our brains can be confused by an illusion which makes two pictures look like a single image - putting a fish inside a fishtank.

Equipment

- Pencil
- White card (white paper glued onto card works too)
- Colouring pens
- Clear tape / sticky tape
- Ruler
- Scissors

Instructions

1. Fold the card in half and use the ruler to draw a 5cm by 5cm (2in x 2 in) square on one side.
2. Cut out the square shape, cutting through both pieces of card so you end up with two identical squares.
3. Mark the centre of each square with the pencil.
4. On one of the pieces of card, use the marker pen to draw a large fishbowl, ensuring its centre is in the centre of the card.
5. On the other piece of card, use the pens to draw a colourful fish, much smaller in size than the bowl, but also with its centre in the centre of the card.
6. Place the two pieces of card together, ensuring the images are on the outside and that they are both the same way up.
7. Use tape to stick the top, left and right edges of card together.
8. Slide the pencil into the centre of the opening at the bottom of the cards, then tape across the opening, securing the pencil in place.
9. Hold the pencil between both palms and quickly spin the pencil by rubbing your hands together.
10. Watch the card when it is spinning. If you spin it fast enough, the two images will look like one and the fish will appear in the tank!

The Science Behind Fish in the Tank

The card and pencil device is called a thaumatrope. When the pencil is spun quickly enough, the two images on the card move faster than our brains can process, so it merges them into one image rather than two separate images. This turns the image of the fish and the image of the fish tank into an animation in which the fish appears to be in the tank. When our eyes see hundreds of still images moving quickly in sequence, it connects them to form a continuous stream of motion - this is how traditional cartoons are made.

Explore Further

» Spin the pencil more slowly and note at what speed your brain can still see the separate images.
» Try drawing different images - such as a bird and a bird cage or a spider and a web.
» What happens if you use larger squares of card, does the experiment still work?

Straw ROCKETS

Become your own rocket designer with this fun experiment that shows how mass, thrust and force are crucial for a successful launch.

Equipment & Ingredients

- Drinking straws
- Paper
- Pencil
- Scissors
- Clear tape / sticky tape
- Ruler

Instructions

1. Cut a strip of paper 3 to 5 cm (1 to 2 in) wide.
2. Trim the strip so that it winds around a pencil once, to form a tube. Use sticky tape to hold the tube shape in place.
3. Remove the tube from the pencil and seal one end by folding over and securing with sticky tape.
4. Use the remaining paper to cut out fin shapes and tape onto the sides of the rocket at the open end of the tube.
5. Place the rocket over the end of a straw and blow hard through the straw to launch it!

The Science Behind the Straw Rockets

All rockets, whether made from paper tubes or carbon fibre, require a few things to fly well. Firstly, they need to have enough force to start moving. When you blow air into the straw, it tries to flow out of the other end. However, because the rocket is blocking the end of the straw, you need to blow hard in order to create enough force to push the rocket off the end so that the air can pass through the straw. The harder you blow, the more energy you provide and the further your rocket will fly.

The shape of the rocket is also important, as air resistance can cause drag and slow it down. Drag is the force of air pressure that pushes against the rocket. Long pointy shapes experience less drag than large round shapes and so will also fly further. Fins stabilise the rocket as it flies through the air, helping to keep it balanced and reducing the chance of it spinning or tumbling, so it can fly further and for longer.

Eventually, gravity will pull your rocket down. The heavier your rocket, the more force will be required to make it fly a given distance, so the trick is to use as little paper and tape as possible while using enough to still hold it all together.

Explore Further

» What happens when you change the flight path of your rocket? Does it travel further if you point it up or straight out? Why do you think that is?

» Add another fin to your rocket. How does this affect how your rocket flies? Why do you think the additional fin does this?

» Do you think you could make a larger rocket that fits over a kitchen roll tube? Would it fly as well as if you blew it through the tube?

Have fun making these
rockets using your favourite
colours and your own designs.

Balloon SHUTTLE

Blast your own speedy balloon shuttle across the room as you learn about Newton's third law of motion.

Equipment & Ingredients

- Long balloon (round ones will also work)
- Drinking straw
- Clear tape / sticky tape
- Scissors
- 3m (approx. 10ft) length of string or cotton

Instructions

1. Tie one end of the string to a chair, door handle, or other support.
2. Cut the straw in half. Feed the free end of the string through one of the pieces of straw.
3. Ask another person to hold the free end of the string, pulling it tight, or tie the string to the top of a chair.
4. Blow up the balloon and pinch the end so that the air stays in - but do not tie it.
5. Stick the side of the balloon to the straw using a piece of sticky tape, so that the balloon hangs underneath the string.
6. Slide the balloon to one end of the string, making sure the mouth of the balloon is closest to the tied end of the string.
7. Let the end of the balloon go and watch it shoot across the room!

The Science Behind the Balloon Shuttle

Rockets work by forcing gas out of a nozzle at high speed which pushes the rest of the rocket in the opposite direction. This is due to Newton's third law which says that for every action there is an equal and opposite reaction. Usually, when you let the air out of a balloon it spins around, shooting off in different directions. By attaching the balloon to the string via the straw, the direction the balloon can travel in is controlled. The thrust creates a forward motion, which pushes the balloon along the direction of the string as the air rushes out.

Explore Further

» Does changing the shape of the balloon, or the amount of air in the balloon change how fast or how far it flies? Why do you think this is? How could you measure the speed of the balloon?

» Change the angle of the string by raising or lowering one end. Does it change the speed that the balloon flies if it is flying uphill or downhill? Why do you think that is?

» What happens if you tape the balloon onto the straw so that the mouth is pointing sideways to the string direction? Does the balloon shuttle still fly?

Bouncing BALL

Bouncy balls are usually made out of rubber, but in this simple experiment you can make and decorate your own bouncy ball.

Equipment & Ingredients

- Microwaveable dish
- Spoon for stirring
- 50ml (10 tsp) warm water
- 28g (3 Tbsp) cornstarch/cornflour
- Food colouring

Instructions

1. Pour the cornflour into the dish, add 25ml (5 tsp) of warm water and mix well using a spoon.
2. Heat the mixture in the microwave for 20 seconds.
3. Add another 25ml of water and 1 drop of food colouring if desired.
4. Mix well using the spoon, then roll the mixture into a ball shape.
5. Place the ball in the microwave and heat for 15 seconds to set the shape.
6. The ball is now ready to bounce. If cracks start to form, dab a few drops of water over the cracks.
7. Keep the ball in an airtight container to stop it drying out.

The Science Behind the Bouncing Ball

Cornflour or cornstarch is made up of tiny grains of starch. When water is added, these grains float around - or suspend - in the fluid. The starch consists of molecules of glucose which is a type of sugar. When heated in the microwave, the starch swells up in the water and some of the bonds between the glucose molecules break. This causes the starch to release some of the glucose molecules into the water, producing a warm gel in a process called gelatinisation. As the mixture cools down, molecules of amylase, which are also in the starch, start to bind together in a molecular mesh structure which holds the solid ball shape together. The heat also causes some of the water to evaporate from the ball, leaving a much higher concentration of starch in the solution, so making a firmer and more solid ball. Although the ball is a solid, it can still be 'squished', meaning it is elastic in its behaviour. When the ball is dropped it becomes compressed when it hits a surface. However, due to its elastic properties, it quickly returns to its original shape, pushing back on the surface as it does so, causing the ball to bounce back up in the air.

Explore Further

» Can you use other flours or starches to make a bouncy ball? What about tapioca or rice flour? Do they bounce as high?

» What other items could you use to decorate your ball? Would adding glitter or poster paint change the height that the ball bounces?

» What happens if you mix 1 part water to 2 parts cornstarch? Will it form a bouncy ball?

» How does this mixture behave if you tap it quickly with the back of a spoon compared to leaving the spoon on the surface of the mixture?

SOLAR Cookie Oven

Usually, ovens are powered using electricity or gas. This experiment uses the sun's own natural energy - or solar power - to heat an oven that bakes cookies.

Equipment & Ingredients

- Pizza box (a cereal box will work)
- Pencil
- Ruler
- Utility knife
- Aluminium foil
- Plastic wrap

- Clear tape / sticky tape or glue
- Black paper
- A warm, sunny day!
- Cookie dough: 15g (1 Tbsp) butter, 30g (2 Tbsp) sugar, 25g (2 Tbsp rounded) plain flour, 5ml (1 tsp) milk

Instructions

1. Use a pencil and ruler to mark 2.5cm (1 inch) in from each side on the pizza box lid and draw a square to connect the marks.
2. Cut along the front and side lines of the square with the utility knife. Fold along the back line so that it acts as a hinge.
3. Line the inside of the flap and the inside walls of the box with foil, and tape or glue in place.
4. Seal the cut-out window shape by taping a sheet of plastic wrap over the hole.
5. Glue or tape a sheet of black paper to the bottom of the inside of the box.
6. Mix the cookie dough ingredients and roll into small balls, then flatten with your hand.
7. Arrange the discs of dough on a sheet of foil and place on top of the black paper inside the box.
8. Prop the cut-out lid at a 75 degree angle (approx) using a pencil and some tape. Leave outside facing the sun until cookies are ready (anything from 15-60 minutes).
9. Carefully remove the cookies from the box and allow to cool - then eat!

The Science Behind the Solar Cookie Oven

Solar ovens use heat emitted from the sun, called solar energy, to cook food. The foil reflects sunlight into the box, and the plastic wrap acts like a greenhouse by preventing the heated air inside the box from escaping. As more heat from the sun streams into the box, the air inside gradually becomes warmer. The black paper in the bottom of the box absorbs the warm sunlight, which in turn heats the food placed on top of it.

Explore Further

» What other items can you cook in the solar oven? Could you bake a potato or even a pizza?

» Does changing the angle of the reflector flap change how efficient the solar oven is at cooking?

» Try using the oven on a warm but cloudy day. Does it still cook the cookies? Why do you think that is?

ERUPTING Volcano

Volcanos are mountains that erupt with molten rock, called lava, when enough pressure builds up. In this experiment you can build your own volcano and watch it erupt again and again.

Equipment & Ingredients

- Small plastic bottle
- Masking tape
- Large tray
- Scissors
- Mixing bowl
- Spoon

- Hairdryer (optional)
- Poster paint and paintbrush (optional)
- Newspaper
- 100g (1 cup) flour
- 200ml (1 cup) cold water

- 200ml (1 cup) vinegar
- 15g (3 tsp) baking soda
- Red food colouring
- 30ml (2 Tbsp) warm water
- Dishwashing liquid

Instructions

1. Place the empty plastic bottle in the centre of the tray and run masking tape from the top of the bottle down to the bottom of the tray to make a volcano-shaped frame.

2. Add the flour to the cold water in a bowl and stir well to make a glue.

3. Tear the newspaper into long thin strips. Soak the strips for a few minutes in the flour-glue mixture.

4. Remove the strips from the glue, one at a time, squeezing off any excess liquid. Lay each strip across the tape frame to build up a volcano layer.

5. Continue to add layers until you have built up a volcano shape. Let each paper layer dry before applying the next one. If you wish, you can use a hairdryer to speed up the drying process between layers. Allow to dry overnight.

6. Paint and decorate your volcano to make it look more realistic.

7. Carefully pour the baking soda into the plastic bottle inside the volcano.

8. Mix 200ml (1 cup) of vinegar with 30ml (2 Tbsp) warm water, 6 drops of dishwashing liquid and two drops of red food colouring.

9. Pour the liquid mixture into the bottle, stand back - and watch your volcano erupt!

The Science Behind the Erupting Volcano

Glue is an adhesive which sticks substances together. By adding water to flour, the molecules in the flour are hydrated, which makes them sticky. When applied to the paper, they help the layers of paper to stick on top of each other and, once the water evaporates from the flour, give rigidity to the paper making it stronger. When an acid such as vinegar is added to an alkali such as baking soda, they react with each other. In this case, the reaction produces the gas, carbon dioxide. The gas becomes trapped in the dishwashing liquid and produces gas filled bubbles. As more and more gas is produced, the pressure inside the bottle increases as there is not enough room for all the bubbles and gas inside the bottle. The gas and bubbles try to release the pressure by rapidly pouring out of the top of the bottle - resulting in a bubble filled eruption.

Explore Further

» What happens if you change the amount of vinegar or baking soda used in the volcano? Does the experiment work with other acids - such as lemon or lime juice - and other alkali materials - such as laundry detergent? Do you get the same type of eruption?

» Can you think of ways to modify the top of the volcano that might increase the pressure inside the bottle to make the eruption go higher or last longer?

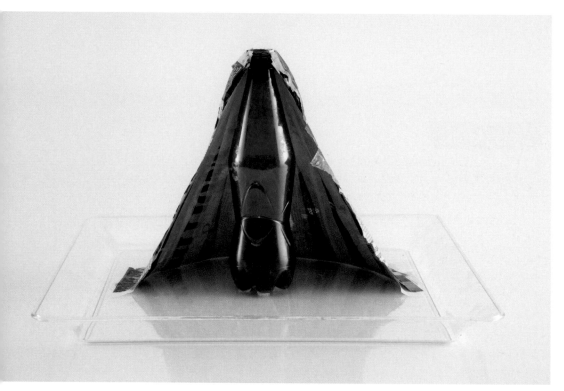

Build your volcano as large or small as you like by
changing the size of the bottle that you use.

Section 3

EDIBLE

EXPERIMENTS

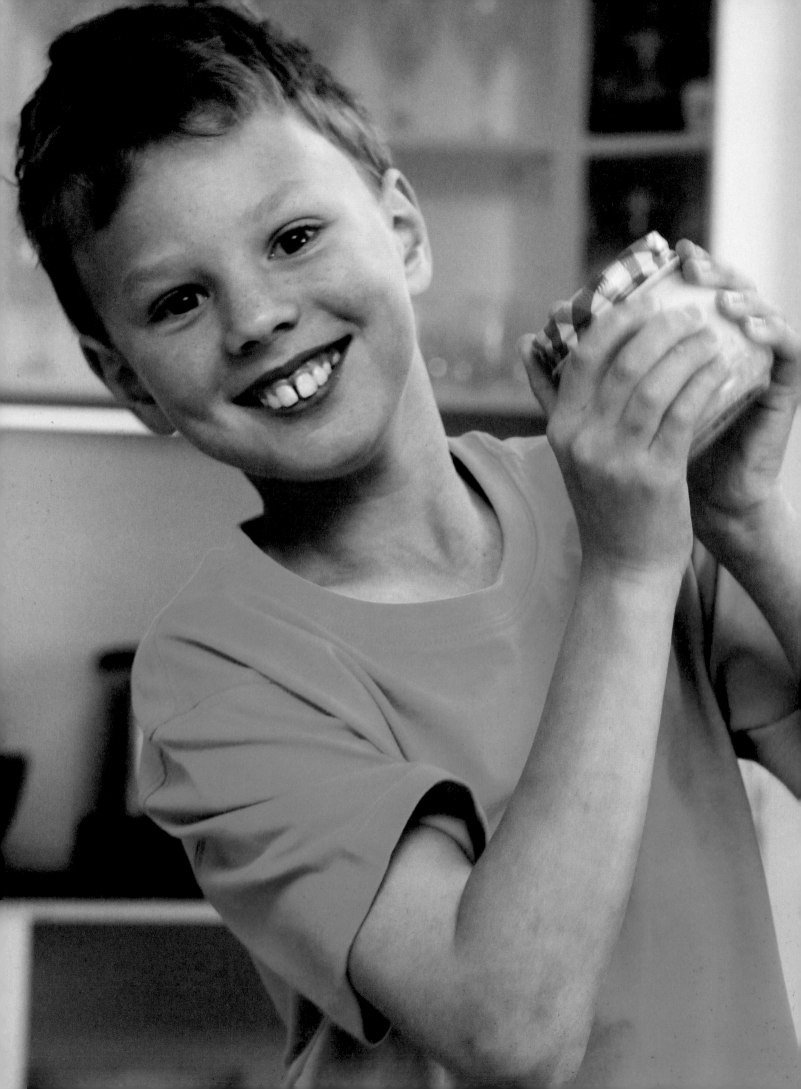

Making
BUTTER

Make fresh butter and give your arms a great workout at the same time with this shakingly good recipe!

Equipment & Ingredients

- Glass jar with lid
- 100ml (1/2 cup) fresh cream/whipping cream
- Cold water

Instructions

1. Pour the cream into the jar, ensuring it fills less than half of the jar.
2. Tighten the lid and turn the jar upside down (over the kitchen sink!) to check the lid does not leak.
3. Vigorously shake the jar for approximately five minutes. You should be able to hear the cream sloshing around inside the jar.
4. Open the jar and see if the cream has thickened and turned into whipped cream.
5. Tighten the lid again and keep shaking the thickened cream for another five to ten minutes or until you hear a runny liquid 'sloshing' sound inside the jar.
6. Open the jar and pour out the clear liquid - this is buttermilk.
7. The yellow solid left inside the jar is butter. Rinse a few times in cold water to wash off any remaining buttermilk and it will be ready to spread on your toast!

The Science Behind Making Butter

Milk is made up of fats and proteins which are suspended in a liquid. Scientifically, this is known as a colloid or an emulsion. If the milk is allowed to sit for a while, the tiny fat particles float to the top, creating a layer of cream which can be scraped off.

The fat particles or globules in the cream are held together by a skin. When the cream is shaken, the fat globules start to clump together, trapping tiny air bubbles in between them. This results in a light and air-filled mixture known as whipped cream.

If the mixture is shaken even more, these air bubbles burst and the skin around the fat globules bursts, causing the fat particles to spill out. Continued shaking causes the newly freed fat particles to join together, forming a solid fat mixture. This is the butter, which has been separated from the liquid in the cream. The liquid is known as buttermilk and is great for making pancakes and breads more fluffy.

Explore Further

» Does the experiment change if the starting temperature of the cream is different? Try very cold cream or cream that is at room temperature.

» What happens if you place a clean, glass marble in the jar to make the butter? Does the butter form more quickly or more slowly? Why do you think that is?

» Place the butter in iced water and squeeze out any remaining buttermilk. This should look like the butter you find in the supermarket, but without any additives it will not taste the same. What flavourings could you add to your butter to make it even more delicious?

Milking NUTS

This tasty experiment mixes the nutritious proteins and oils from almonds into a drinkable liquid, full of energy, that looks just like milk.

Equipment & Ingredients

- Blender
- Sieve (fine mesh)
- Bowl

- 200g (1 1/2 cup) almonds
- 950ml (4 cups) water

Instructions

1. Soak the almonds overnight in a bowl of water.
2. Rinse and peel off the skins.
3. Place the peeled almonds in a blender with 950ml (4 cups) water.
4. Blend until smooth.
5. Pour through a fine mesh sieve, squeezing to remove all the liquid.
6. Refrigerate the liquid until chilled, then drink!

The Science Behind Milking Nuts

Cow's milk is a mixture of nutrients, proteins, fats and sugars, produced to feed baby calves and help them grow. Almonds are actually the seeds of the almond tree. They contain a high concentration of proteins and fats to nourish young plants as they grow into trees.

When the nuts are soaked and blended in water, the oils and proteins are released from the solid nut into the liquid. Because oil and water do not mix, the oil from the almonds forms tiny droplets which sit in the liquid forming an emulsion. Almond milk is not technically a milk. However, because it looks like cow's milk, and contains similar fat and protein levels, it is often referred to as a milk.

Explore Further

» What other nuts could you blend and turn into a milk?
» Experiment with adding other flavours like vanilla essence or cinnamon to enhance the milk.
» Dry out the almond pulp that is left over to form almond meal which can be used to make gluten free cakes or macarons.

IRON in your Breakfast

'Fortified with Iron' is commonly seen on many food items, but what does iron in your food actually look like? This experiment will show you how to find the iron in your breakfast cereal.

Equipment & Ingredients

- Rolling pin
- Zip-sealing food bag
- Magnet (a fridge magnet will do but a stronger one is preferable)
- White sheet of paper
- 25g (1 cup) breakfast cereal labelled 'fortified with iron'
- 100ml (1/2 cup) warm water

Instructions

1. Pour the cereal into the food bag and seal.
2. Using the rolling pin, crush the cereal in the bag until it becomes a fine powder.
3. Pour the powder onto the paper in a thin layer.
4. Run the magnet closely over the top of the cereal powder and look for any small black particles being picked up by the magnet.
5. Return the cereal powder to the bag, add 100ml of warm water, and reseal.
6. Put the magnet on the table, and place the bag on top of the magnet.
7. Gently swirl the fluid around in the bag then leave for 15 minutes.
8. Keeping the magnet in contact with the bag, gently turn the bag over and look carefully at the inside of the bag close to where the magnet is.
9. You should be able to see a small cluster of black particles around your magnet. This is the magnetic iron from your cereal which is being attracted to the magnet.

The Science Behind Iron in Your Breakfast

Magnetism - ferromagnetism in this case - is a complex and interesting phenomenon which is behind many electric motors and generators. Ferromagnetic properties are caused by tiny electrons which spin around atoms having their own magnetic fields. When these tiny fields align they can cause an attractive force, such as that found when iron is close to a magnet.

Iron helps your blood to carry oxygen molecules from your lungs to the rest of your body. Because we cannot produce iron naturally, we need to obtain it from our food by eating meats and nuts. Many food makers add iron to their foods to help us consume enough to remain healthy - these foods are 'fortified with iron'. In breakfast cereals, the iron is compressed into the cereal structure. When the cereal is crushed into a fine powder, the structure is broken up, allowing the iron particles to fall out of the cereal. Soaking the cereal powder in water helps to dissolve the cereal, so that the tiny iron particles can further separate from the cereal structure and move around in the water. When the magnet is placed close to the watery cereal, the magnetic attraction force is enough to pull the iron through the water and towards the magnet. If enough iron collects around the magnet, it can be seen with the naked eye as a clump of tiny black dots.

Explore Further

» See if you can identify which brands of cereals contain the most iron. Do your findings correlate with the nutritional list on the side of the cereal box?

» What other foods can you find that are fortified with iron?

» Not all metals are magnetic. Can you use your magnet to find other magnetic metals around the house?

Confectionery CANDLE

Many candles are thrown away after being used. However, with this yummy recipe you can have your candle and eat it!

Equipment & Ingredients

- Matches
- Knife
- Plate

- Banana
- Almond
- Chocolate or nut pieces (optional)

Instructions

1. Peel the banana and cut the ends off to make a flat-based banana cylinder.
2. Stand the banana upright on the plate and decorate with chocolate or nut pieces.
3. Peel the skin off the almond and carefully cut lengthways to make a thin slice.
4. Push the almond slice into the top of the banana.
5. Use the matches to light the almond and watch it burn.
6. After the experiment, blow out the flame and eat the whole candle!

The Science Behind the Confectionery Candles

Candles are made of two things, a wick and a wax.

The wax acts as a fuel, providing energy to the flame, allowing the candle to burn. Because the wax is usually solid in a candle, the heat from the flame softens it until it becomes a liquid. The wick then absorbs the liquid wax and pulls it up towards the flame. When the liquid wax reaches the flame it turns into a vapour or a gas which fuels the flame, keeping it burning.

For a flame to continue to burn it needs fuel, energy and oxygen. The wax provides the fuel, the initial energy comes from the match, and the candle is surrounded by oxygen in the air.

The edible candle also has initial energy from the match and has oxygen surrounding it. Instead of using wax, the edible candle uses the almond as both a wick and a fuel. Nuts are high in energy because they are filled with natural fats. These fats burn slowly and, when lit, provide fuel for the flame. The banana acts as a base for the almond to sit in, and its high moisture content keeps the flame safe and minimises the risk of the fire spreading.

Explore Further

» Do other types of nuts - such as cashews or walnuts - work as good wicks? Which type of nut burns for the longest? What do you think that tells you about the nut?

» Try other solid foods as a base - such as an apple or an orange - instead of a banana, if you prefer the taste of those.

» Does the thickness of the sliced nut change how easy it is to light? Why do you think that is?

Nuts are full of energy which can provide the fuel needed for a flame to burn

Unicorn NOODLES

These amazing and edible unicorn noodles can transform from purple to blue or pink right in front of your eyes.

Equipment & Ingredients

- Large saucepan
- Knife
- Stove
- Large heatproof bowl
- Sieve or colander

- Red Cabbage
- Lemon
- Clear noodles (vermicelli or glass noodles work well)
- Hot water

Instructions

1. Roughly chop the purple cabbage leaves and place in the saucepan.
2. Add enough water to the saucepan to half cover the cabbage leaves.
3. Bring to the boil and cook for 5 minutes on the stove.
4. Place a colander over a large heatproof bowl and strain the hot cabbage.
5. Put the cabbage aside - if you like, you can add a pinch of salt and dash of vinegar to make it into a tasty side dish!
6. Pour the cabbage juice back into the pan and add the noodles.
7. Simmer for 5-10 minutes, until the noodles are soft and purple.
8. Use the colander to drain off the water and transfer the noodles to a plate or bowl.
9. Squeeze fresh lemon juice onto the noodles and watch them turn pink!

The Science Behind Unicorn Noodles

Purple cabbage is purple due to a pigment called anthocyanin. This same pigment is also found in blueberries. As the cabbage boils, the anthocyanin leaches out into the water. When the dehydrated noodles are added to the cabbage water the anthocyanin is absorbed. Scientists use a scale called the pH Scale to describe the concentration of hydrogen protons in a solution. With 7 being neutral, a pH of less than 7 means the solution is acidic whilst a pH greater than 7 means the solution is alkaline. Anthocyanin changes colour depending on the pH of the solution it is exposed to. When it is neutral (or at pH 7) it is purple, but if it comes into contact with something acidic such as lemon juice, it turns pink. An alkaline solution, on the other hand, would make the anthocyanin turn blue, green or even yellow. In addition to being a tasty snack, the unicorn noodles are also an edible pH meter!

Explore Further

» What happens when you sprinkle an alkaline material such as baking soda onto the noodles?

» Can you estimate the pH value of other household products - such as vinegar or laundry powder - using the leftover cabbage juice?

» Using what you now know about anthocyanins, can you explain why the blueberries in blueberry muffins sometimes look green around the edges?

Microwave CHEESE

This quick and tasty experiment will have delicious ricotta cheese on your plate in under 20 minutes!

Equipment & Ingredients

- Sieve
- Paper towels
- Heatproof jug
- Bowl

- 1 cup full fat milk
- 1g (1/4 tsp) salt
- 15ml (1 Tbsp) white vinegar

Instructions

1. Pour the milk, salt and vinegar into the jug and stir.
2. Heat in the microwave for 3 minutes or until it starts to bubble around the edges.
3. Remove from the microwave and stir gently, watching the milk separate into solid, white curds.
4. If the solids do not form, return to the microwave and repeat.
5. Line the sieve with 4 paper towel sheets and place over a bowl.
6. Spoon the milk mixture into the sieve to drain off the liquid.
7. Tip the strained solid into a bowl and it is ready to eat, either as it is, or spread on toast.
8. To create sweet ricotta, add a dash of honey or sprinkle of sugar to the mixture.

The Science Behind Microwave Cheese

Milk is made up of fats, sugars and proteins. One of the types of protein in milk is called casein and caseins do not dissolve well. Instead they curl up into tiny spherical structures called micelles which float around in the liquid milk. Liquids that contain small particles like proteins which do not dissolve are called colloids. Other common food colloids are mayonnaise and butter. Adding the vinegar reduces the pH of the milk, making it more acidic. This causes the spherical micelle structures to denature and uncoil. The heat speeds up the process. As they stretch out and uncoil, the casein proteins become tangled with each other and clump together or coagulate, forming a solid structure known as curd. The liquid that is left over is called whey. (Curds and whey are mentioned in the nursery rhyme 'Little Miss Muffet'!) The final drained cheese produced is more commonly known as ricotta and is often used in Italian cooking.

Explore Further

» Does the experiment work with different types of milk such as raw milk, low-fat milk, UHT milk or soy milk? Why do you think some milks work better than others?

» Can you find other edible acids instead of vinegar which also break down the casein proteins - such as lemon juice or buttermilk?
Does this change the taste of the ricotta cheese?

» Experiment with adding different flavours to your microwave cheese - which do you like the most?

INSTANT
Ice Cream

Scientific Principle
PHASE CHANGE

10 MINUTES

This delicious science recipe will give you ice cream in under 10 minutes!

Equipment & Ingredients

- One small zip-resealable sandwich bag
- One large zip-resealable plastic bag
- 120ml (1/2 cup) cream or full fat milk.
- 12.5g (1 Tbsp) sugar
- A few drops of vanilla or other flavouring of your choice
- 3-7 cups of ice
- 75g (5 Tbsp) salt

Instructions

1. Add the cream, sugar, and vanilla to the small bag and seal, ensuring that any excess air is released.
2. Place the ice, salt and cream-filled bag into the larger bag and seal.
3. Vigorously shake the large bag over a sink for approximately 5 minutes. Stop when the cream has started to freeze and turn into a solid.
4. Remove the small bag and quickly rinse off the salt solution with cold water.
5. Pour the ice cream into a bowl, add your favourite toppings and enjoy eating your newly frozen dessert!

The Science Behind the Instant Ice Cream

Ice cream is an emulsion, or a mixture of two liquids (water and fats) which do not normally mix together. To make ice cream, the milk or cream mixture needs to change its state from a liquid to a solid. If the mixture was simply placed straight into the freezer, the water component would freeze first, forming large, crunchy ice crystals. Ice cream tastes better when it is creamy rather than crunchy, so the goal in ice cream making is to create the smallest ice crystals possible. By vigorously shaking the bag, any large ice crystals that may be forming are broken up into smaller crystals, resulting in a smooth and creamy ice cream. The freezing point of ice is lowered by the addition of salt, so it starts to melt. As this ice melts it draws heat energy from its surroundings - including the cream mixture enclosed in the smaller bag - cooling it enough to cause the liquid cream emulsion to freeze, changing it from a liquid to a solid and forming ice cream.

Explore Further

» What happens if you do not shake the bag vigorously when making the ice cream?
» If you put too much ice cream in your mouth, you may suffer from what is called 'brain freeze' or an 'ice cream headache'. Placing your tongue on the roof of your mouth should stop the headache - why do you think this is?
» Taste the ice cream frozen, then taste it again when it has melted. One should taste much sweeter than the other - why do you think this is?

Candy CRYSTALS

Watch in awe as you grow your own edible, crunchy, candy crystals - the longer you leave them, the bigger they grow!

Equipment & Ingredients

- Wooden skewer
- Clothes peg
- Saucepan
- Tall, narrow, clean glass or jar
- 1 cup of water
- 2-3 cups of sugar
- Food colouring

Instructions

1. Heat the water in a saucepan over a low heat until it is simmering.
2. Slowly add the sugar, stirring constantly, making sure that the sugar dissolves in the water before adding more.
3. Keep adding the sugar until the water starts to look cloudy. This is the point where no more sugar will dissolve.
4. Remove the pan from the heat and allow to cool.
5. Wet skewers with water, then roll them in the remaining sugar - leave for a few minutes to dry.
6. Once the sugar solution has cooled, pour it into the glass or jar and add food colouring.
7. Clip the clothes peg onto the wooden skewer and suspend with the peg over the top of the glass so that the skewer is in the centre of the glass and approximately 2cm (1 in) from the bottom of the glass. Leave the glass on a table where it will not be disturbed.
8. The first crystals should form after 3 days and will continue to grow.
9. You can help your candy crystals to grow by checking for, and removing, any crusty film that forms on the surface of the solution.
10. When you are happy with the size of your candy crystals, remove from the solution and allow to dry for a couple of hours before eating.

The Science Behind Candy Crystals

If you pour a spoonful of sugar into a glass of cold water and stir, the sugar will dissolve into the water. Eventually, if you keep adding sugar to the water it will stop dissolving. However, if the water is heated, more sugar can be forced to dissolve in the water, creating what is called a supersaturated solution. As the water cools, the supersaturated solution becomes unstable since it contains more sugar than it can hold. The sugar then starts to come out of the solution and reforms as solid sugar crystals.

When the sugar starts to come out of the solution it finds the lowest energy surface to form on. As it takes less energy for the sugar crystals to form on top of other crystals than to form on their own in the solution, the sugar-rolled skewers act as seeds for the new sugar crystals to grow. The more the sugar solution cools, and the more water ev6porates from the solution over time, the more sugar crystals come out of the solution - and the bigger the candy crystal grows.

Explore Further

» Can you think of ways to flavour your candy crystals such as with peppermint oil or vanilla essence? Do you think this will change the structure of your candy crystals?

» Can you make crystals with other crystal-forming materials such as salt? Do the crystals look the same or different?

» How big can your candy crystal grow? Will it keep growing forever or eventually reach a maximum size? Why do you think that is?

When sucrose molecules are cooled slowly they arrange
in a regular pattern creating a sugar crystal.

EDIBLE Earthworms

Scientific Principle
CROSS-LINKING

45 MINUTES

(PLUS 4 HOURS SETTING TIME)

Using the power of crosslinking, these realistic looking worms not only look amazingly disgusting - they also taste great!

Equipment & Ingredients

- 50 bendy drinking straws
- Elastic / rubber band (or length of string for tying)
- Jug
- Tall container
- Plate

- 2 boxes jelly/jello strawberry or raspberry crystals or blocks
- 10g (1 Tbsp) powdered gelatin
- 125ml (1/2 cup) cream
- 375ml (1 1/2 cups) boiling water
- Green food colouring

Instructions

1. Carefully pour the boiling water into a large jug and add the jelly/jello and the gelatin, stirring until dissolved.
2. Add the cream and whisk until fully mixed.
3. Stir in 3 drops of green food colouring.
4. Stretch the flexible part of the straws out so that they are fully extended.
5. Gather the straws together and use a rubber band or string to hold them together.
6. Place the straws upright in a tall, tight fitting container or jar.
7. Carefully pour the mixture over the top of the straws, filling each straw. Refrigerate for 4 hours.
8. If the straws start to float, place a weight on top to hold them down.
9. Once set, rinse the outside of the straws in lukewarm water to loosen the worms.
10. Starting at the top, gently squeeze each straw together with your fingers (or the back of a blunt knife) and slide down the length of the straw to push the worms out onto a plate.
11. To make the worms look as though they are in soil, crush dark chocolate cookies and lay them on the plate as a base for the worms to sit on.

The Science Behind Edible Earthworms

Jelly/jello wobbles because it contains gelatin, a coiled up protein chain that unravels and floats around as strands when hot water is added. As the water cools down, the gelatin strands coil back up and become tangled with each other, trapping the fluid they are in and transforming the liquid into a solid structure. This process of gelatin strands becoming tangled with each other is called crosslinking. Because the worms have a high aspect ratio - meaning they are long and thin - they need to be stiffened to help them keep their shape. Adding the extra gelatin causes more crosslinking to occur, with the chains making the structure firmer and stiffer when set. Jelly/jello is usually transparent - or see through - but the additional protein and fat molecules from the cream deflect and scatter the light so that the worms become opaque. Mixing red jelly/jello and green food colouring makes the worms a 'realistic' brown colour - but you can, of course, make them any colour you like.

Explore further

» What happens if you change the amount of gelatin in the worm mixture? Why do you think this is?

» Do the worms look different if you do not add cream to the mixture?

» What happens if you do not rinse the straws in warm water before squeezing out the worms? Why do you think the warm water helps?

The elastic properties of these edible earthworms means
that if they are moved, energy easily transfers from one side
of the worm to the other which makes them wobble.

This slime flows like a liquid but can be rolled like a solid - and the best part is that it's edible!

Equipment & Ingredients

- Saucepan
- Plastic sandwich bag
- 395g (14 oz) can sweetened condensed milk
- 10g (1 Tbsp) cornflour/cornstarch
- 45ml (3 Tbsp) chocolate syrup

Instructions

1. Pour the milk into the saucepan and heat on a low heat.
2. Slowly stir the cornflour into the warm milk. Continue heating and stirring over a low heat for 20 minutes or until the mixture thickens.
3. Remove from the heat and stir in the chocolate syrup.
4. Place in a sandwich bag and refrigerate.
5. Once cold, roll and squeeze into any slimy shape you want - and watch it flow!

The Science Behind Scrumptious Slime

Cornflour or cornstarch is a starch made up of long chains of sugar molecules called glucose which are joined together in a coiled up ball. When exposed to heat and milk, the starch particles absorb water from the milk, causing them to swell. These swollen particles start to press up against each other. This reduces the movement of the liquid, resulting in it thickening or becoming more viscous. Eventually the starch particles burst, freeing up long strands of starch which swell further and absorb the fluid outside the particles. This traps the remaining water in the mixture and turns it into a highly viscous gel or slime. The slime flows like a thick liquid but can be rolled around like a soft solid. The advantage of this recipe is that the slime is edible once you have finished with it!

Explore Further

» What edible treats could you add to your slime to add more texture? Does this change the way that it flows?

» How does the slime flow differently when it is warm compared to when it is cold? Why is this?

» Can you think of other ingredients you could add, instead of the chocolate sauce, to make different flavoured edible slime?

The colder the temperature
of the slime, the less runny
or more viscous it becomes.

Honey COMB

Watch as this chemical reaction expands before your eyes to make a delicious treat that is both sweet and bubbly!

Equipment & Ingredients

- Baking Tray
- Parchment or baking paper
- Saucepan
- Sieve

- 1 1/2 cups white sugar
- 1/2 cup honey
- 1 Tbsp baking soda (sifted)
- Pinch of salt

Instructions

1. Line a baking tray with parchment or baking paper.
2. Pour the honey, salt and sugar into a saucepan and heat on a medium to high heat, stirring continuously.
3. Continue to heat for 3 minutes or until the mixture turns an autumn brown colour.
4. Remove from the heat and add the baking soda, stirring as it foams.
5. Quickly pour the mixture onto the baking tray and leave to cool for 10 minutes.
6. Peel from the paper and break into bite-sized pieces.

The Science Behind Honeycomb

Baking soda is usually added to cakes and breads to help them rise. In this experiment, the heat of the sugar solution causes the baking soda to break down and release carbon dioxide. This release of gas into the sugar solution causes it to bubble and produce an expanding foam. Spreading out the foaming mixture onto the large baking tray results in a thin layer which has a large surface area. This allows the mixture to cool very quickly and solidify the bubbles in place, creating a rigid honeycomb structure. Because of this quick cooling process, the sugar molecules do not have time to arrange themselves into sugar crystals and so form a glassy amorphous or non-crystalline structure, which makes the honeycomb brittle and easy to break.

Explore Further

» What would happen if you added double the amount of baking soda to this recipe? Do you think you would get double the bubbles?

» How would the honeycomb differ if you poured it into a bowl instead of onto a tray? Why do you think it would change the properties of the final structure?

» Break off a piece of honeycomb and look at the bubble structure inside. Are the bubbles the same size and shape? Why do you think that is?

BREAD
in a bag

This no-mess recipe makes delicious fresh bread, thanks to the fermentation power of yeast - with the advantage of there being no bowls to wash!

Equipment & Ingredients

- Large zip-sealing plastic bag
- Lightly oiled loaf pan
- Kitchen towel
- 360g (3 cups) white flour
- 35g (4 Tbsp) sugar

- 7g (2 1/2 tsp) yeast
- 5g (1 tsp) salt
- 45ml (3 Tbsp) oil
- 240ml (1 cup) warm water

Instructions

1. Preheat the oven to 190° Celsius, 375° Fahrenheit.
2. Pour the sugar, yeast and 1 cup of flour into the bag and seal. Shake to mix the dry ingredients.
3. Open the bag and pour in the warm water. Let as much air out of the bag as possible then reseal the bag.
4. Squeeze and scrunch the ingredients in the bag until they are well mixed.
5. Let the bag sit for 10 minutes to allow the yeast to start fermenting.
6. Open the bag and add 1 cup of flour and the oil and salt. Reseal and mix the ingredients.
7. Open the bag and add the final cup of flour, reseal and and mix for a further 5 minutes or until the mixture stops sticking to the inside of the bag.
8. Transfer the mixture into a loaf pan, cover with a clean kitchen towel and leave to rise for 30 minutes.
9. Bake for 25 minutes in the oven, then carefully turn onto the kitchen towel and allow to cool for 10 minutes.

The Science Behind Bread in a Bag

When flour is mixed with water the starch in the flour absorbs the water and it binds to form a sticky, cohesive solid. The yeast, which is a single-celled fungus lies dormant until it comes into contact with warm water. Once reactivated, the yeast starts to feed on the sugars. You will notice a sour odour from the alcohol produced by this fermenting yeast if you open the bag at this point. Carbon dioxide gas is produced as a byproduct of fermentation and this becomes trapped as gas bubbles in the dough, which causes the bread to rise.

When the dough is heated in the oven, the gas pockets expand making the bread rise further and the alcohol evaporate. The higher temperature also kills the yeast and causes the sugars on the outside of the bread to caramelise creating a nice a brown crust.

Explore Further

» Yeast is crucial to helping the bread to rise by creating the gas, carbon dioxide. You can see how much gas is produced by mixing the water, sugar and yeast (from the recipe for this experiment) into a bottle, and sealing the top of the bottle with a balloon. What happens to the balloon over time? Do the results change if the water used is very cold or very hot? Why do you think that is?

» What do you think would happen if the sugar was removed from the bread ingredients? Can you think of other ingredients you could add to the bread to make it more tasty?

Fun FOAMSICLES

Create a tasty, fluffy treat on a stick while learning about the delicate science of whipping egg whites.

Equipment & Ingredients

- Popsicle or lollipop sticks
- Baking tray
- Parchment or baking paper
- Glass or steel mixing bowl

- Metal whisk
- 100g (½ cup) white sugar
- 2 egg whites
- 5ml (1 tsp) lemon juice

Instructions

1. Preheat oven to 95°Celsius, 200°Farenheit.
2. Line a baking tray with baking paper and lie the wooden sticks approximately 10cm (4 in) apart on top of the paper.
3. Whisk the egg whites until you can form soft peaks with the mixture.
4. Whisk in the lemon juice. Whisk in the sugar a little at a time.
5. Continue to whisk until you can form stiff peaks with the mixture.
6. Spoon the mixture into round drops on one end of each of the wooden sticks.
7. Bake for 1 hour 30 minutes. Remove from the heat and leave to cool before eating.

The Science Behind Fun Foamsicles

Egg whites contain water and proteins which are made up of long chains of amino acids. These proteins are normally curled up, but whisking adds air bubbles to them, causing them to uncurl and stretch out. Parts of the uncurled protein like water (hydrophilic) and parts dislike water (hydrophobic). When they stretch out, the hydrophobic parts surround the air bubbles in order to stay dry, holding the bubbles in place. Fats and oils can disrupt the stability of these protein coated air bubbles which is why any unseparated egg yolk or traces of oil in the bowl will reduce the foaminess of the whisked egg white.

Lemon juice is an acid. When you add an acid to a mixture, you are essentially adding some positively charged particles. These positively charged particles are hydrogen ions—hydrogen atoms that have lost an electron. The hydrogen ions hop onto charged portions of the proteins and leave them uncharged. This helps to stop too many of the proteins from bonding with each other which would otherwise make the egg white mixture lumpy and cause it to collapse.

Once exposed to heat, the gas inside the air bubbles expands, making the bubbles bigger. The heat also makes the proteins solidify around the expanding bubbles, creating a solid foam.

Explore Further

» Try whisking the egg whites using first a cold egg and then an egg which is at room temperature. How does the starting temperature affect the final structure of the egg white foam? Why do you think this is?

» What happens to the foam structure if you keep whisking the egg whites? Can you over-whisk them? Why do you think the foam structure changes with too much whisking?

» Time how long it takes to make a foam structure if the sugar is added to the egg whites before whisking. Why do you think the whisking time for egg whites is different if sugar is added compared to when it is not?

Section 4

ELECTRICITY

EXPERIMENTS

the Static Powered DANCING ghost

Watch in amazement as your tissue ghost magically dances without you touching it, thanks to an invisible electrostatic force.

Equipment & Ingredients

- 2-ply or 4-ply tissue
- Balloon
- Scissors
- Pen
- Clear tape / sticky tape

Instructions

1. Peel apart each of the tissue layers so you end up with a very thin, single ply sheet.
2. Using the pen, draw a 'ghost' shape approximately 4cm (1.5 in) tall.
3. Cut out the shape using scissors and add any features you want to with a pen.
4. Tape the base of the tissue ghost to a table.
5. Blow up and tie off the balloon.
6. Rub the balloon across your hair for around 10 seconds.
7. Bring the balloon close to the ghost and watch the ghost rise up towards the balloon, appearing to fly on its own. Woo hoo!

The Science behind the Static Powered Dancing Ghost

When some objects are rubbed together, they can create a buildup of electrical charge known as static electricity. Rubbing a balloon across your hair causes the negatively charged electrons to transfer from your hair onto the balloon. This results in a negatively charged surface on the balloon. The tissue sheet is neutral in charge, which means that it has equal amounts of both positive and negative charges. However, because opposites attract, when the negatively charged balloon comes close to the tissue, the positive charges within the tissue become attracted to the negative charges in the balloon and rearrange to move towards the balloon. The force of attraction is enough to pull the lightweight tissue sheet up towards the balloon and the ghost shape is picked up by the static charge.

Explore Further

» Does rubbing the balloon back and forth change the amount of static charge compared to only rubbing in one direction?

» If you rub the balloon against your hair for a longer amount of time, does it produce more static charge?

» Try rubbing with materials other than your hair - such as wool or silk fabric. Do they create more or less static charge?

Plasma GRAPES

After solids, liquids and gas, plasma is the fourth state of matter. Usually only seen in the laboratory, this experiment lets you create your own plasma in a flash!

Equipment & Ingredients

- Microwave
- Knife
- Small microwave-safe plate
- Paper towels
- Grapes

Instructions

1. Cut a grape in half lengthways, taking care to leave a small amount of skin intact so that the halves are still held together.
2. Open up the halves and place the grape, cut side up, on a small microwave-safe plate.
3. Pat the cut sides of the grape with a piece of paper towel to dry off any excess moisture on the surface.
4. Place the plate inside the microwave and cook on high for 5 seconds.
5. Watch as a flash of brightly coloured flame forms above the grape.
6. Be careful when removing the cooked grape from the microwave as it will still be hot.

The Science Behind Plasma Grapes

Microwaves heat food using a type of wave similar to radio waves and light waves. They work because microwaves are absorbed by water, fats and sugars, causing them to heat up. Grapes contain lots of liquid, which is why they are nice and juicy when you eat them. The liquid in a grape is an electrolyte, which means that the fluid contains ions or atoms with a positive or negative charge. The size of a grape is much smaller than the 12cm (5 in) wavelength of the microwave. This means that the grape can act as an antenna, focusing the microwave power into the middle of the grape. As the microwave heats the grape, the charges in the electrolytes start to travel back and forth very quickly between the layer of skin connecting the two halves of the grape. Eventually this 'skin bridge' heats up so much that it dries out and the charges become trapped. To release this charge, the ions jump or arc through the air from one half of the grape to the other, creating a bright ion cloud, filled with highly excited electrons.

Explore Further

» What happens if you cut the grape in half and do not connect the two pieces together? Does the plasma still form? Why do you think this is?
» Why do you think that you should not put conductive materials like metals in the microwave?
» What happens when you heat an air-filled material - such as a marshmallow - in the microwave for 30 seconds? Why do you think this happens?

Levitating RING

Although it might look like 'magic', this impressive levitation experiment is pure electrostatic science!

Equipment & Ingredients

- Lightweight plastic bag (often used in supermarkets for fruit and vegetables)
- Balloon
- Woollen clothing or scarf
- Scissors

Instructions

1. Cut a narrow (1cm or about 1/2 in) strip across the width of the closed bag and open to form a ring.
2. Blow up and tie the balloon.
3. Rub the woollen item several times across both the surface of the balloon and the plastic strip to build up a static charge on both materials.
4. Throw the ring up in the air, away from your body, with enough force to prevent it sticking to your hand.
5. Hold the balloon underneath the ring as it falls to the ground.
6. Watch as the plastic ring levitates above the balloon.

The Science Behind the Levitating Ring

When certain materials are rubbed together they create what is called static electricity. Static electricity is the collection of electrically charged particles called electrons on the surface of a material. The electricity is called 'static' because the electrons stay in one area rather than flowing or moving to another area. You may already have experienced a static shock - perhaps if you wear natural fabric clothing in the winter. The shock comes from your clothing rubbing as you move - a static charge is built up which is eventually released when you, for example, touch another person or get out of the car.

Natural materials - such as wool, hair or fur - tend to leave other materials positively charged when rubbed against them. Some materials - such as rubber and polyethylene - tend to leave other materials negatively charged when rubbed against them. Opposite charges attract each other, whereas similar charges repel or move away from each other. Rubbing the wool fabric against both the balloon and the plastic strip results in the electrons creating a static charge on the surface of both the balloon and the plastic. Because the balloon and the plastic strip now both have a negative static charge on their surface, they repel by creating a force which pushes them away from each other. Since the plastic strip is very light, the repulsion force created is strong enough to push it away from the balloon. The balloon appears to hold the plastic strip in the air and causes it to levitate - the magic of science!

Explore Further

» What happens if you don't fully inflate the balloon in this experiment?
» Why do you think the the plastic strip wants to stick to your hands after it has been rubbed with the wool?
» Can you successfully repeat the experiment, replacing the wool fabric with other natural materials - such as your hair or a piece of silk?
» Does the experiment work better or not so well in a humid bathroom?

Section 5

MOTION

EXPERIMENTS

Whirlpool
IN A BOTTLE

Create your own swirling whirlpool using a bottle, some arm strength and the power of centripetal force.

Equipment & Ingredients

· Transparent bottle or jar with a lid or cap.
· Water
· Dishwashing liquid

Instructions

1. Fill the bottle or jar with water until it is 3/4 full.
2. Add a few drops of dishwashing liquid.
3. Tightly screw on the lid and turn the bottle or jar upside down over a sink to make sure that it does not leak.
4. Keeping the bottle upside down, hold it with both hands and vigorously swirl for 5 seconds.
5. Stop swirling and look inside the bottle or jar. You should see a whirlpool forming in the centre.
6. You might need to practise your spinning technique a few times to make a bigger whirlpool.

The Science Behind Whirlpool in a Bottle

When liquid spins in a circular motion it can form a whirlpool. You may have seen this effect in the bathtub when you pull out the plug. This is caused by the increased speed of the water as it passes through the small opening in the plug hole. Whirlpools commonly occur in nature when water is first forced through a narrow opening and then flows into a more open area. Spinning the bottle or jar in a circular motion causes the water to spin rapidly around its centre, due to centripetal force. Centripetal force is a force that pulls the water towards the centre of its circular path which, in this case, is the centre of the bottle. When the water is poured out of the bottle, it will make a spiral pattern called a vortex. This is caused by the force of gravity pulling the liquid down out of the bottle while the spinning water still inside the bottle rotates around a centre point as it tries to pour out.

Explore Further

» Add glitter or food colouring to the water - does it help you to see the whirlpool better?
» Does the speed that you spin the bottle change the size of the whirlpool formed?
» Change the amount of water you place in the bottle - does more water make bigger whirlpools?
» Use your hand to cover the bottle opening instead of a lid and repeat the experiment over the sink. Remove your hand to let the water flow out - what happens?

Hoop DROP

Amaze your friends by dropping a coin from a height directly into a small mouthed bottle - thanks to the science of inertia.

Equipment & Ingredients

- Small coin
- Stiff paper or card
- Small food jar or bottle with a mouth slightly larger than the coin
- Pen
- Scissors
- Clear tape / sticky tape

Instructions

1. Cut the card into a long strip about 2cm (1 in) wide and about 25cm (10 in) long and connect the ends with tape to form a hoop.
2. Secure the jar on a level surface and place the hoop vertically on top of the mouth of the jar.
3. Balance the coin on the top of the hoop, directly above the mouth of the jar.
4. Hold the pen through the centre of the hoop, and in one quick horizontal motion, swipe the hoop off the jar.
5. If you do this quickly enough, the hoop should fly out of the way, and the coin will drop directly into the jar below.

The Science Behind Hoop Drop

Newton's first law of motion describes how an object at rest will remain at rest unless acted upon by an external force. By swiping the hoop quickly out of the way, there is not enough friction to keep the hoop and coin in contact, so the hoop slides out from under the coin. Without a hoop to support it, the coin is pulled straight down by the force of gravity and drops into the jar below it.

Explore Further

» Does changing the size of the hoop change the experiment? Why do you think that is?

» Can you balance multiple coins on top of each other on the hoop? Do they all drop into the jar?

» Instead of swiping from the centre, place the pen outside the hoop and swipe it off horizontally. Does the coin still drop into the jar? Why do you think this is?

This experiment uses the same principle that stops you falling out of a spiralling roller coaster to keep water inside a spinning cup!

Equipment & Ingredients

- Disposable cup
- String
- Scissors
- Water

Instructions

1. Carefully pierce two holes opposite each other, near the top rim of the cup.
2. Cut a piece of string about as long as you are tall (up to a maximum of 150cm/60 in).
3. Tightly tie each end of the string to the opposite holes in the cup to form a long, looped handle.
4. It is not advisable to try this experiment indoors! Please ensure you go outside to an open area where water can be spilled without causing any damage.
5. Half fill the cup with water.
6. Hold the string tightly in the centre and gently start swinging the cup from left to right.
7. Once you have developed confidence in swinging the cup, use the same motion to spin it in a circle over your head several times.
8. If you spin the cup fast enough, you should find that the water stays in the cup even when it is upside down.

The Science Behind Spinning Water

Usually, when a cup filled with water is turned upside down, the water pours out.
This is because the force of gravity pulls the water down towards the Earth. However, in this experiment, the water stays in the cup.

Newton's first law states that objects in motion will remain in motion unless acted on by some external force - the principle of inertia. If you let go of the string as you spin the cup, the cup will fly off in a straight line. With the string attached, the cup will keep spinning in a circular path. The reason why the cup does not fly away is, of course, because it is tied to your hand! The string applies a centripetal force to the cup - a force that always pulls towards the centre of the circle, and keeps the cup travelling in its circular path. The centripetal force pulls the cup and water back towards the centre (where your hand is) and stops it flying off. If this force is strong enough, it can overcome the force of gravity on the water and prevent the water from pouring out - even when it is upside down! For this experiment to work, the cup must travel fast enough for the centripetal force to remain stronger than the effect of gravity. If you swing it too slowly, the water will fall out... and you will get wet!

Explore Further

» What happens if you change the length of the string attached to the cup?
» Does the experiment still work if you swing the cup horizontally above your head?
» Will the experiment still work if you use a bucket of water and rope instead?

Section 6

PRESSURE

EXPERIMENTS

Potato JACKHAMMER

A drinking straw is usually only used for sucking up liquids, but in this experiment a straw will be transformed into a cutting tool, using only the power of air pressure.

Equipment & Ingredients

- Stiff drinking straws
- 1 uncooked potato

Instructions

1. Take a potato and hold it by its sides, ensuring the end of the potato is left exposed.
2. Pick up a straw with your other hand.
3. Holding the straw by its sides, see if you can stab it all the way through the potato.
4. Repeat the experiment with a new straw but, this time, place your thumb over the top of the straw so that the hole is covered.
5. With your thumb over the top of the straw, you should be able to stab the straw straight through the potato!

The Science Behind Potato Jackhammer

The first time you tried the experiment, the straw may only have penetrated a short way into the potato before the straw bent. This is because straws are made of thin plastic which is not very strong.

The real secret to this experiment is the air inside the straw. By placing your thumb over the hole at the top, the air is trapped inside inside the straw. This trapped air forces the air molecules to compress or get closer together because they are unable to escape from the top of the straw. The compressed air inside the straw gives the straw strength by stopping the sides from bending as it hits the potato.

You may have seen road workers using jackhammers to dig holes in the road. Jackhammers work in the same way, where the force that makes a jackhammer pound up and down is provided by an air hose. The force of the air pressure makes a hole in the road, just like the hole in the potato.

Explore Further

» Does the length of the straw make a difference to how well it can cut through the potato?
» Can you use the straw to cut through other food items like an apple or a pear?
» Look at your thumb after the experiment - can you see a shape? Why do you think this is?

Usually when a glass of water is turned upside down, the water pours out. However, with a piece of cardboard and the power of air pressure, a glass can be turned upside down with no spillage of water at all!

Equipment & Ingredients

- A drinking glass
- A piece of cardboard
- Scissors
- Water

Instructions

1. Cut the cardboard into a piece slightly larger than the diameter of the glass.
2. Fill the glass to the top with water.
3. While holding the glass steady with one hand, use your other hand to place the cardboard over the mouth of the glass.
4. While gently supporting the cardboard with the palm of your hand, turn the glass upside down. You might want to do this over a sink - just in case!
5. Gently remove your hand from the cardboard. It should defy gravity by sticking to the glass and preventing the water from pouring out.

The Science Behind Weightless Water

The force of gravity usually pulls things down. Without the cardboard, the water would pour out of the glass. However, placing the cardboard on top of the glass stops any more air from getting into the glass. When the glass is turned upside down, a small pocket of air can be seen at the top. This is an area of low pressure, formed because the pocket does not contain many air molecules in the small space at the top of the glass. The air molecules outside the glass want to move into the glass to try and make the air pressure the same inside as outside. As more air molecules push up against the bottom of the card, the air pressure pushing up on the cardboard is increased. The force is enough to hold the cardboard in place, even when the water is pushing down on it due to gravity.

Explore Further

» If the glass is not filled all the way to the top, the air space in the glass will have a greater density of molecules. This air space can be large enough so that the air pressure is equal to the air pressure outside of the glass.

» Experiment by pouring different amounts of water into the glass and observe what happens.

» Does the thickness of the cardboard make a difference to the experiment?

» What happens when you cut a small hole in the centre of the cardboard? Why do you think this is?

Egg
IN A BOTTLE

Watch in amazement as a solid egg squeezes into a bottle, using the power of heat and air pressure.

Equipment & Ingredients

- Small candles
- Glass jar or bottle with a mouth slightly smaller than the egg
- Long matches
- Saucepan
- Water
- Egg
- Banana

Instructions

1. Place the egg in a pan and cover with cold water. Bring the water to the boil, then lower the heat and simmer for 8 minutes.
2. Carefully remove the egg and rinse in cold water. When the egg is cool, peel off the shell.
3. Peel a banana, and cut a slice 2cm (1 in) in length to use as a candle holder.
4. Place the candle in the centre of the slice of banana and lower into the bottom of the glass jar.
5. Use a long match to light the candle.
6. Place the egg on top of the mouth of the bottle to seal it.
7. Watch as the flame goes out, and the egg slowly gets sucked inside!

The Science Behind Egg in a Bottle

Without the candle, when the egg is placed on top of the bottle, the air pressure inside the bottle would match the air pressure outside the bottle, and nothing would happen. When the candle is alight inside the bottle, it causes the air around it to heat up and expand. Placing the egg over the mouth of the bottle then seals the bottle and prevents any more air from getting inside. The candle flame needs oxygen to burn. Once all the oxygen inside the sealed bottle is used up, the flame goes out. Without a flame heating the air, the warm air inside the bottle starts to cool down. Cool air takes up less space than hot air, so as the air inside the bottle cools it contracts, lowering the air pressure inside the bottle.

The air pressure outside the bottle is now greater than the air pressure inside the bottle, so the air outside starts to push down onto the bottle and the egg. Eventually the egg is pushed into the bottle, releasing the seal and balancing the air pressure inside and outside the bottle.

To remove the egg from the bottle, the air pressure inside the bottle needs to be increased. This can be done by turning the bottle upside down and tilting it until the small end of the egg is sitting in the open neck of the bottle. Now place your mouth on the end of the bottle and blow. This forces more air into the bottle, raising the pressure inside. When you take your mouth away, the egg should pop out!

Explore Further

» What happens if you light two candles in the bottle? Does the egg squeeze into the bottle twice as fast?

» Does the experiment take a longer or shorter time to work if a softer boiled egg is used?

» Can you get the experiment to work by heating the air inside the bottle in a different way - such as using a hairdryer?

Air pressure is the force of air molecules pushing against things

and can be powerful enough to squeeze an egg into a bottle!

Watch as a metal can crushes right in front of your eyes - using only the force of air pressure!

Equipment & Ingredients

- 1 empty soft drink can
- Tongs
- Stovetop or camping stove
- Bowl
- Water

Instructions

1. Fill a large bowl with cold water and place it aside.
2. Pour 15ml (1 Tbsp) of cold water into the empty can.
3. Using the tongs to hold the can, heat the base of the can on the stove for approximately 45 seconds.
4. Once steam starts to rise from the top of the can and bubbling can be heard inside the can, quickly turn the can upside down and dunk it into the bowl of cold water, ensuring the open top is immersed.
5. The can should crush as soon as it hits the water.

The Science Behind Can Crusher

Water is liquid at room temperature and pressure but transforms to water vapour as it is heated and boils. Water vapour or steam takes up more space than liquid water, so some of the air inside the can is pushed out to make space for the water vapour. When the can is dunked into the cold water, the water vapour condenses and returns back to its liquid state, which takes up less volume in the can. There is now less air in the can and, because the opening of the can is sealed underwater, air is unable to flow in and fill it back up again. This makes the air pressure inside the can lower than the air pressure outside the can. The higher air pressure outside the can exerts a force onto the can. If the pressure difference is big enough, the force is strong enough to crush the can.

Explore Further

» Do you think the can will still crush if you do not pour water into the can before heating?
» What happens if you chill the cold water further by adding ice cubes? Why do you think this happens?
» Does the size and shape of the drink can affect the extent to which it crushes?

Fireproof BALLOON

Usually a balloon bursts when you bring a flame close to it, but thanks to the heat-absorbing properties of water, this experiment can bring the two together without a bang.

Equipment & Ingredients

- Candle
- 2 balloons
- Tongs
- Matches or a lighter
- Water

Instructions

1. Blow up and tie one of the balloons.
2. Light a candle and use tongs to hold the balloon 1cm (1/2 in) above the flame. WARNING! the balloon is likely to burst within a few seconds!
3. Fill the other deflated balloon with water, then carefully blow it up (ensuring the water stays inside the balloon), before tying the end.
4. Repeat the experiment, using tongs to hold the balloon 1cm (1/2 in) above the flame, and see what happens.
5. If you are very brave, you can try lowering the balloon further so that the flame touches the bottom of the balloon!

The Science Behind Fireproof Balloon

The heat from a flame normally bursts a balloon, as you would have seen with the air filled balloon. This is because the heat thins the balloon rubber until it creates a hole. The internal air pressure inside the balloon forces the hole to grow outwards quickly, splitting the balloon material. Water is very good at absorbing heat energy, so when the water filled balloon is held over the flame, the water closest to the flame - rather than the rubber - absorbs the heat. This warm water rises up, and cooler water moves in to replace it at the bottom of the balloon. The constant rising of warm water and flowing in of cool water keeps the bottom of the balloon cool and prevents the rubber material of the balloon from heating up and thinning out. As long as there is cold water available in the balloon to replace the warm water, a constant exchange will prevent the balloon from popping, even if it is placed in the flame. Only when the heat of the flame is greater than the water's ability to conduct heat away from the thin rubber, will the balloon pop - but this may take several minutes, depending on how much water is in the balloon.

Explore Further

» Do you think the experiment will work if you fill the balloon with hot water?
» Will it make a difference if the balloon is only half filled with air instead of fully inflated?

Section 7

REACTION

EXPERIMENTS

Raising RAISINS

Raisins usually sink in water, but with this gas-producing chemical reaction, you can use bubble power to help them float to the top.

Equipment & Ingredients

- 1 transparent glass
- Tablespoon
- Warm water
- 5 raisins

- 10g (2 tsp) baking soda/bicarbonate of soda
- White vinegar

Instructions

1. Half fill the glass with warm water.
2. Add the raisins to the water and note what happens.
3. Add the baking soda to the water and stir gently.
4. Watch what happens to the raisins.
5. Fill the glass to the top with white vinegar and watch the raisins - they should rise and fall in the glass.

The Science Behind Raising Raisins

Density is a property that determines if something will float or sink. Raisins are more dense than water, so they sink when they are placed in the water-filled glass. Vinegar is acidic and baking soda is alkaline. When vinegar and baking soda are mixed together, they react to form a gas called carbon dioxide which is released as small bubbles. Bubbles are less dense than water, so they rise to the top of the glass. Because raisins have ridges and folds, some of the bubbles become trapped in these spaces. This increases how much space or volume the raisin/bubble takes up in the water. If enough bubbles become trapped in the ridges of the raisin, the combined raisin/bubble unit becomes less dense than water and the raisin floats up. When the raisin reaches the top, the bubbles burst and the raisin goes back to being more dense than the water so it sinks to the bottom again. The science of bubble power!

Explore Further

» Why does this experiment use warm water? Would it work with cold water?

» What happens when the bubbles run out? Is there a way to make more bubbles?

» Does the experiment work with other bubble-filled liquids - such as fizzy soda drinks?

Bouncy EGGS

Usually when you drop an egg it cracks, but in this experiment you can drop an egg and watch it bounce!

Equipment & Ingredients

- Glass or jar
- Egg
- White vinegar
- Water

Instructions

1. Place the egg in the glass and add white vinegar until the egg is completely submerged.
2. Cover the top of the glass or seal the jar to prevent the vinegar evaporating.
3. Leave the egg in the vinegar for 4 days, observing any changes in appearance.
4. Remove the egg from the glass and rinse under water, carefully rubbing the shell from the egg.
5. Gently squeeze the egg and note how rubbery it feels.
6. Hold the egg around 5cm (2 in) above a table top - you should see the egg bounce like a bouncy ball!

The Science Behind Bouncy Eggs

The shell of an egg is made of calcium carbonate which is brittle and cracks easily. When the egg is submerged in the vinegar, bubbles of carbon dioxide form on the surface of the eggshell. These are produced by the acid of the vinegar reacting with - and dissolving - the calcium carbonate in the egg shell. Once the shell has dissolved, the egg is no longer protected from the vinegar and it starts to absorb it. The egg is made up of proteins which can be denatured (or its natural qualities changed) when exposed to heat or acid. This denaturing can also be seen when a raw egg is cooked and it goes from being runny and transparent to solid and white. Vinegar can do the same thing, transforming the outside of the egg into a translucent, rubbery solid which is tough enough to be dropped without breaking.

Explore Further

» What happens if you use a boiled egg instead of a raw egg?
» How high can you drop the egg before it bursts? Does it look like a normal egg inside?
» Does the egg change in size after it has been in the vinegar?
» What other liquids do you think the egg could be immersed in to achieve the same effect?

The shell of an egg is made of calcium carbonate which is brittle and cracks easily. When the egg is submerged in vinegar, bubbles of carbon dioxide form on the surface of the eggshell. These are produced by the acid of the vinegar reacting with – and dissolving – the calcium carbonate in the egg shell.

Rubber BONE

Tie a bone in a knot with this fun experiment and learn why calcium is so important for strong bones.

Equipment & Ingredients

- Glass jar with lid
- Chicken leg bone
- Vinegar

Instructions

1. Rinse the bone in running water, scraping off any residual meat.
2. Try to bend the bone by holding at both ends and note how difficult it is.
3. Place the bone in the jar, cover with vinegar, and seal the jar.
4. Leave in the jar for 7 days.
5. Remove the bone and rinse in water.
6. Try to bend the bone again, and see how much easier it is than before.

The Science Behind Rubber Bone

Bones are made up of a mixture of hard minerals and tough, flexible proteins like collagen. It is this combination which makes bones not only strong, but also able to flex without shattering. Calcium is a mineral which helps to give bone its strength. This is why babies need a lot of calcium-rich milk to strengthen their bones as they grow. Vinegar is an acidic liquid which is strong enough to dissolve the calcium out of the bone and make it less stiff. With more collagen than calcium, the bone goes from feeling hard and stiff to soft and rubbery.

The relationship between structure and properties is very important to engineers and is essential in the design of structures that are both strong and lightweight, just like bone.

Explore Further

» Does the experiment work with other acids - such as lemon juice or fizzy soft drink?
» Do you think the size of the bone affects the time taken for the calcium to dissolve?
» Would immersing the bone in a calcium rich fluid - such as milk - make it stonger?

FLOATING Eggs

Scientific Principle
DENSITY

10 MINUTES

Density determines how well things float. You may have noticed that you float more in the ocean than in a swimming pool. This is due to the salt in the ocean making the water more dense. This experiment shows how you can change the density of water and help a sinking egg to float.

Equipment & Ingredients

- 1 tall transparent glass
- Tablespoon
- Water
- 1 egg
- Salt

Instructions

1. Place the egg into the glass.
2. Fill the glass with water, ensuring that the egg is completely submerged.
3. Note if the egg floats or sinks.
4. Carefully remove the egg from the water using the spoon.
5. Add 2 Tbsp of salt to the water in the glass and stir until completely dissolved.
6. Place the egg back into the glass and observe to see if the egg floats or sinks.
7. If the egg behaves in the same way as it did in Step 3, remove the egg and stir in 2 Tbsp more salt.
8. Repeat, noting how much salt is needed until the egg floats.

The Science Behind Floating Eggs

Density is a measure of how much stuff or matter is in a certain amount of space or volume. The more matter that can be packed into a given volume, the more dense it is.

When salt is added to water, it makes the water more dense, meaning that the same volume of water has more matter in it. Put simply, the water becomes heavier from the additional weight of the dissolved salt in it, without the water taking up any more space. The more salt is added, the heavier or more dense the water becomes.

Objects float when they are less dense than the fluid they are in.

At the beginning of the experiment, the egg is more dense than the water so it sinks. As salt is added to the water, it increases in density. Eventually the point is reached at which the water becomes more dense than the egg and so the egg floats.

Ocean water contains salt and is more dense than fresh water. This is why it is easier for a person to float in the ocean than in a swimming pool.

Explore Further

- » What else can you dissolve in the water to make the egg float?
- » Are there other objects that already float in water? Why might that be?
- » Does the water look any different when you stir in salt?
- » Does the temperature of the water make any difference to how much salt is needed to make the egg float?

At the beginning of the
experiment, the egg is more dense
than the water so it sinks.

As salt is added to the water, it increases in density. Eventually the point is reached at which the water becomes more dense than the egg and so the egg floats.

Sticky ICE

In cold climates, salt is put on roads to help melt dangerous ice. This experiment uses the same principle to melt ice cubes in a way that lets you hold them with a piece of string!

Equipment & Ingredients

- 1 plate
- 1 piece of string cut to about 30cm (12 in) long
- Timer
- Ice cubes
- 5g (1 tsp) salt

Instructions

1. Place a cube of ice in the middle of the plate.
2. Lay the string across the top of the ice cube and wait for 30 seconds.
3. Lift the string up off the ice cube. Did it stick?
4. Lay the string back across the ice cube.
5. Sprinkle the salt over the top of the ice cube and wait for 30 seconds.
6. Hold both ends of the string and lift upwards. The ice should be stuck to the string and you should be able to pick it up off the plate.

The Science Behind Sticky Ice

Usually, ice melts and water freezes at 0° Celsius or 32° Fahrenheit. However, salt lowers the temperature at which ice can melt and water can freeze. By adding the salt to the ice cube, the ice around the string starts to melt, and the freshly melted water moves up on top of the string and re-freezes. This traps the string in a layer of freshly frozen ice, enabling it to be picked up off the plate.

Pressure can also make ice melt at colder temperatures. This is how ice skates glide in ice rinks. The pressure from the blade melts the ice when the blade pushes down on it, helping the blade to glide easily on a thin layer of water, and so enabling the ice skater to skate!

Explore Further

» Can you lift more than one ice cube onto a piece of string?
» If you have access to different types of salt - such as rock salt or sea salt - do they work as well?
» What if you use other string-like materials - such as an elastic/rubber band or a strand of hair?

Rest your cheeks and inflate your own balloons using the science of gas production

Equipment & Ingredients

- Small plastic bottle
- Balloon
- Sheet of paper
- Clear tape / sticky tape

- 120ml (1/2 cup) vinegar
- 15g (3 tsp) baking soda

Instructions

1. Pour the vinegar into the bottle.
2. Blow up and deflate the balloon a few times to stretch it out.
3. Make a funnel shape by rolling the paper so it has a small hole at one end and a large hole at the other end. Secure with tape.
4. Have one person stretch the end of the balloon open while another person uses the funnel to pour the baking soda into the balloon.
5. Twist the end of the balloon to seal it so that the baking soda is trapped inside the balloon. Then carefully stretch the open end of the balloon over the neck of the bottle, taking care that the baking soda does not drop into the bottle.
6. Position the balloon (with baking soda inside) so that it hangs over the side of the bottle.
7. Tape the bottle to a solid surface so that it cannot fall over, or have one person hold the bottle steady.
8. When you are ready, lift the balloon up so that the baking soda falls from inside the balloon into the vinegar-filled bottle.
9. Watch as the mixture produces bubbles and the balloon starts to inflate.

The Science Behind Bubbly Inflator

When vinegar and baking soda are mixed together, a chemical reaction occurs which produces a gas called carbon dioxide. A chemical reaction is where two or more substances interact to create new substances, or products. In this reaction, the vinegar (an acid) reacts with the baking soda (an alkali) to produce carbon dioxide gas, water, and a third chemical called sodium acetate.

The gas formed through this reaction takes up much more space - more volume - than the liquid and solid we began with. As a result, the gas pushes out of the bottle and starts to fill the space inside the balloon. As more gas is created, the balloon inflates, increasing the space inside the balloon.

When the balloon is blown up up before the experiment, the polymer material it is made from is stretched out, making it easier to inflate the balloon with just the small amount of air pressure formed by the chemical reaction inside the bottle.

Explore Further

» What happens if you don't pre-inflate the balloon before putting baking soda inside it?
» If you double the amount of baking soda and vinegar, does the balloon double in size?
» Does the experiment work with other acidic liquids - such as lemon juice?
» Does swirling the bottle to mix the baking soda and vinegar affect the balloon size?

When vinegar and baking soda are mixed together,
a chemical reaction occurs which produces a gas called carbon dioxide.

Section 8

SOUND
EXPERIMENTS

CHICKEN in a cup

Pretend that there is a chicken in the room with this funny and easy-to-make instrument that sounds like a chicken in a cup.

Equipment & Ingredients

- Plastic drinking cup
- 40cm (16in) wool, yarn or cotton string
- Wooden skewer
- Paper towel
- Scissors
- Water

Instructions

1. Carefully punch a hole in the centre of the bottom of the cup, large enough for the string to fit through.
2. Tie one end of the string to the middle of the skewer.
3. Push the other end of the string through the hole from the outside of the cup and pull it through the centre of the cup.
4. Fold a paper towel into quarters, then dip it into the water and squeeze out any excess.
5. Holding the cup firmly in one hand, wrap the damp paper towel around the string, close to the mouth of the cup.
6. Squeeze the string with the paper towel and pull down, using short jerks, so that the paper towel slides down the string.
7. Listen carefully and each jerk motion should sound like the cluck of a chicken.

The Science Behind Chicken in a Cup

The sounds that we hear with our ears are created by sound waves. Sound waves are vibrations in the air. When the wet tissue is pulled across the string, the friction - or sliding force of the movement - creates vibrations. These vibrations on the string are too small for our ears to pick up so usually we are unable to hear them. The cup picks up the vibrations and amplifies them, making them loud enough to be detected by the hairs in our ears, and so allowing us to hear the sound.

This theory of amplification is also used by instruments such as pianos and guitars which are made from wood. The wood acts as a sounding board which makes the instrument louder.

Explore Further

» What happens if you use a different size of cup?
» Does the experiment work if the paper towel is dry?
» If you change the string material, does the sound change?

Ringing (((STRING)))

This sound experiment will let you hear a secret gong, which nobody else around you will be able to hear, thanks to the science of how sound waves pass through solids.

Equipment & Ingredients

- Two metal spoons
- 120cm (approx. 48 in) string

Instructions

1. Fold the string in half to find the middle.
2. Tie the middle of the string around the top of the handle of one of the metal spoons and secure with a knot.
3. Wrap one end of the string around one of your index fingers, and the other end around your other index finger, so that the spoon dangles in the middle.
4. Ask somebody to tap the hanging spoon with another metal spoon. Listen to the sound produced.
5. Put your fingers holding the string against each ear, as if you are plugging your ears from noise, making sure the spoon is hanging freely at your waist.
6. Ask somebody to gently tap the hanging metal spoon with the other metal spoon. Listen to the sound coming through the string. Does it sound different?

The Science Behind Ringing String

When the two spoons hit each other they vibrate, creating sound waves. These sound waves travel both through the air and directly along the string to your ears. Because sound vibrations travel better through a solid than through the air, the waves, which travel up the string to your ears, are much easier to hear than the waves which reach your ears by passing through the air.

The sound that you hear will change depending on the size of the metal spoon used. Different sized spoons create different wavelengths of sound due to the atoms that make up the spoons vibrating differently. The volume of the sound will also change depending on how hard the spoon is tapped. The harder it is hit, the greater the amplitude or size of the sound wave that is created, and so the louder the volume of sound.

Changing the length of the string between the spoon and your ears will also change the sound you hear. The faster an object vibrates, the higher the pitch of the sound. Shorter string lengths vibrate more quickly than longer string lengths, so the spoon noise will have a higher pitch as you shorten the length of the string, and a lower pitch when you lengthen it.

Explore Further

» What happens to the sound when you use a wooden spoon to tap the metal spoon? Or two wooden spoons? Why do you think the spoon material changes the sound?

» Shorten the string by wrapping it around your fingers more and listen to the sound produced by tapping two spoons. Does it sound different? Why do you think this is?

» Is the sound the same if you use a metal fork instead of a metal spoon?

» Have fun using the science of sound to pass on 'secret' messages. Attach a plastic cup to each end of the string. Keeping the string taut, listen with the cup to your ear as someone else speaks quietly into the other cup.

Musical STRAWS

Make your own musical instrument using just a drinking straw and the power of your lungs!

Equipment & Ingredients

· Drinking straw
· Scissors
· Pen
· Ruler

Instructions

1. Flatten the straw at one end by squashing the sides together.
2. Measure 2cm (just under 1 in) down from the top of the straw and mark with a pen.
3. Starting at the pen mark, trim the corners off the top of the straw with the scissors to make a pointed end, (see illustration).
4. Separating the two points so they are not stuck together, purse your lips and place them around the 2cm mark of the straw where you started your cut.
5. Blow through the straw.
6. If you don't hear a sound, experiment with moving your lips further away from the end of the straw and vary how hard you blow.
7. Once you have made a sound, experiment by cutting small pieces off the end of the straw to make it shorter and see if the sound changes.

The Science Behind Musical Straws

We hear sound because vibration travels through the air as sound waves. Trimming the end of the straw into two long, thin flaps helps them to vibrate when air is passed through them. As the ends of the straw vibrate, they cause the air inside the straw to vibrate. This vibration creates sound waves that are detected by the hairs inside our ears. The sound produced through the straw is a bit like a duck call. Longer straws produce a lower pitched sound because longer sound waves produce lower notes. By shortening the length of the end of the straw, you can change the pitch of the sound produced.

Explore Further

» What happens to the sound if you cut a longer (3cm or 1 1/2 in) or shorter (1cm or 1/2 in) point in the end of your straw?
» How do you think this changes the way the flaps vibrate?
» Does the straw make a sound if you blow through the other end of the straw? If not, then why not?
» If you cut small holes in the middle of the straw, does it change the sound? What happens when you cover these holes with your fingers and blow through the straw? Why do you think holes in the straw might change the sound?

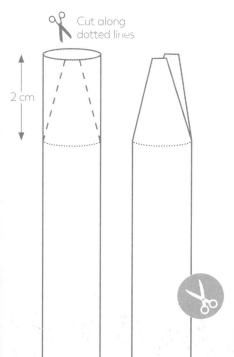

Cut along dotted lines

2 cm

Section 9

SURFACTANT

EXPERIMENTS

Soap powered BOAT

Watch in amazement as a floating boat, with no moving parts, speeds across the water using only the power of surface tension.

Equipment & Ingredients

- Polystyrene foam tray or cardboard
- Shallow tray / dish
- Scissors
- Toothpick
- Water
- Dishwashing liquid

Instructions

1. Fill a tray with clean water.
2. Cut the foam tray or cardboard into a boat shape as shown.
3. Dip the toothpick into the dishwashing liquid and use it to apply a layer onto the sides of the notch at the back of the boat.
4. Carefully place the boat onto the surface of the water and watch it speed across the water.
5. If you want to repeat the experiment, replace the water in the tray with fresh, clean water.

The Science Behind Soap Powered Boat

Surface tension is a phenomenon where the water molecules at the surface of a layer of water are attracted to each other in a way that creates a strong, flexible skin. It is this surface tension that lets the lightweight boat float on the surface of the water. Surface tension also helps insects walk on the surface of a pond.

Dishwashing liquids and soaps are surfactants which means that they break down the surface tension of the water by disrupting the arrangement of the water molecules. As the surface tension breaks behind the boat, the water molecules move from areas of low surface tension to areas of high surface tension. This creates enough force to move the boat across the water towards the areas of high surface tension in front of it. This is known as the Marangoni effect.

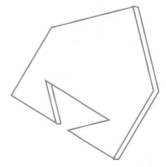

Explore Further

» Does the boat travel faster or slower if the water is warm instead of cold? Why do you think this is?

» Will the experiment still get the boat to move if you use hand soap instead of dishwashing liquid?

» Why do you think the experiment does not work if you repeat it using the same tray of water?

Watch how a beautiful, moving piece of 'food art' changes - right in front of your eyes - thanks to the power of dishwashing liquid which breaks up the surface tension of milk.

Equipment & Ingredients

- Plate
- Cotton bud
- Milk
- Dishwashing liquid
- Liquid food colouring (two or more different colours)

Instructions

1. Pour enough milk onto the plate to cover the bottom.
2. Carefully add a few drops of food colouring onto the surface of the milk.
3. Repeat using different colours, creating a polka dot effect on the surface of the milk. Note how the food colouring floats.
4. Dip a cotton bud into the dishwashing liquid, then hold it in the centre of the plate, keeping it as still as you can.
5. Watch as the food colouring swirls around, creating moving art.
6. Dip the cotton bud into the dishwashing liquid again and touch it onto a different area of the plate.
7. Repeat until the colours stop swirling.

The Science Behind Marbled Milk

Milk is mostly made up of water, but it also contains proteins and fat. Because oil and water do not mix, the fat is stored as tiny droplets which float in the milk. The poured milk holds itself together on the plate using a property called surface tension, which is where the cohesive or sticking forces of the molecules in the milk stick together.

When drops of food colouring are added to milk, they can be seen to float on the surface rather than sinking to the bottom. This is because food colouring is less dense than milk. Dishwashing liquid, designed to break up the grease and fat on dishes to clean them, can also break up fat molecules in milk. The dishwashing liquid breaks the surface tension. The tension across the surface pulls the milk surface away from the break caused by the soapy cotton bud – a bit like when a balloon bursts. As the food colouring is floating on top of the milk it moves with the surface, floating away from the drop of soap and allowing the flow pathways in the milk to be seen. As the soap becomes evenly mixed with the milk, the flow slows down - but all it takes is the addition of another drop of soap to start the process again.

Explore Further

» What happens if you use other dairy products, such as cream, with more fat content?
» Is the flow pattern the same if you use two drops of dishwashing liquid in different places on the milk?
» Does the temperature of the milk change the flow rate, and if so why?

Food colouring floats on top of the
surface of the milk rather than sinking
to the bottom because it is less dense.

Make bath time much more fun by washing with your personally designed, and scientifically crosslinked, wobbly shower cakes!

Equipment & Ingredients

- Silicone muffin tin or small round plastic container.
- 3g (1 tsp) gelatin
- 60ml (1/4 cup) boiling water
- 85g (1/2 cup) bar soap
- 1 drop food colouring
- 1 drop olive oil

Instructions

1. Carefully pour the boiling water into a bowl and add the gelatin, stirring until completely dissolved.
2. Add the oil and food colouring.
3. Grate the soap and add to the bowl, stirring gently to mix.
4. Carefully pour the mixture into a silicone muffin tin or plastic container.
5. Refrigerate for 3 hours.
6. Use in your next shower.
7. Store in an airtight container to stop the shower cake from evaporating.

The Science Behind Shower Cakes

Gelatin is made up of coiled up proteins which are unique because they can be heated up to 100 degrees Celsius without breaking down or denaturing. When gelatin is dissolved in hot water, the protein chains unravel and stretch out, floating around in the water. As the water cools, the gelatin strands start to coil up again, but they become tangled with other gelatin chains, trapping some of the water inside their coiled structures as they do so, through weak bonds called hydrogen bonds. This process is called crosslinking. The coiling up results in the soapy water getting caught in between the molecules so it cannot move freely anymore and the mixture transforms from a liquid to a gel. As you rub the gel on your skin in the shower, the friction or rubbing forces help to release the soap and water from the gel onto your body.

Explore Further

» What happens when you mix more or less of the gelatin into the hot water? How does the structure of the shower cake change? Why do you think this is?

» Can you add other ingredients - such as glitter or peppermint oil - to your cakes to add some sparkle and scent?

» What happens if you do not store your shower cake in an airtight container? Why do you think this happens?

Soap molecules have one
hydrophobic end that is
attracted to dirt and oil and
one hydrophilic end which
protrudes, waiting to be
rinsed away by water.

Everybody loves bubbles, but they usually burst when you try to catch them. This experiment uses sugar and socks to help you bounce bubbles and catch them with your hands!

Equipment & Ingredients

- Small bowl for mixing
- Spoon for stirring
- Cotton (or wool) socks or gloves
- Drinking straw

- 60ml (4 Tbsp) Water
- 30ml (2 Tbsp) Sugar
- 15ml (1 Tbsp) Dishwashing liquid

Instructions

1. Mix all the ingredients together in the bowl and stir until the sugar dissolves.
2. Dip the end of the straw into the solution until a film forms on the end.
3. Gently blow into the other end of the straw to form a bubble.
4. Cover your hand with a sock or glove and hold your palm out flat.
5. Blow a bubble in the air and use your covered hand to gently bounce the bubble without it bursting.

The Science Behind Bouncing Bubbles

Bubbles are simply air trapped inside a thin film of liquid - the bigger the bubble, the more air is inside. The liquid film that makes up the outside of a bubble is mostly water. Water molecules are attracted to each other by intermolecular forces - these are electromagnetic forces which act between molecules. The intermolecular forces draw the water molecules together, creating something called surface tension.

Dishwashing liquid lowers the surface tension of the water, making it stretchy enough to stretch around a sphere of air to form a bubble. Bubbles, though, are prone to popping when the film is pierced or if too much water in the film evaporates, leaving too thin a layer. The sugar binds to the water molecules. This helps to stop the bubbles from drying out so they last longer and don't pop as quickly.

Usually if you touch a bubble it bursts - this is because the natural oil on your hands breaks the surface tension of the water around the bubble. By wearing gloves or socks on your hands you create a barrier between the oil and the bubble, making it possible to bounce a bubble without it bursting.

Explore Further

» Try storing the bubble solution in the fridge overnight. Does it change the size of the bubbles you can blow?
» Try bending the straw into a triangular shape. Can you blow a triangular bubble from it?
» What happens if you add more sugar or more dishwashing liquid - does the quality of the bubbles change?
» Do you think using hot water to make the bubble solution would work?

Thanks...

*At Nanogirl Labs, we believe
that science should be for everyone...*

We dedicate ourselves to projects that we believe in; projects that inspire, educate and empower through science, technology, engineering and mathematics.

From the moment Michelle presented her initial idea for The Kitchen Science Cookbook, our team was captivated and together set to work evolving and developing the book. We have had the very real pleasure of working with a great many wonderful people as we have brought this project to fruition, and I am grateful for the opportunity to recognise and thank those who have played a key role:

Paul Davis, Magic Rabbit Ltd - Photography & Production Management

Val Davis, Magic Rabbit Ltd - Editor

Quent and Jo Pfiszter, Suburban Creative Ltd - Graphic Design

Our wonderful staff at Nanogirl Labs who have worked so tirelessly on this project:

Pauli Sosa, Janet Van, Gabriela Campos Balzat, Katherine Blackburn.

Our volunteer support team, running experiments and supporting the photoshoot:

Gaspar Zaragoza, Parie Malhotra.

A remarkable group of friends and experienced advisors who have so willingly given their expertise, inspiration and encouragement through the course of this, our first publishing project:

Murray Thom, Wendy Nixon and the Thom Productions team. Martin Bell, James Hurman, Robin Ince, Ashlee Vance, Nisha Vasavda, Haley Chamberlain Nelson

The late Raewyn Davies, publicist extraordinaire and a truly special lady. Raewyn's expertise and passion helped us launch this, our first book, with great success. Raewyn taught us so much - and only some of it was about launching books!

Thanks also to our wonderful testing community around the world, to the incredible Kickstarter community, and the 1,193 backers who chose to support this project.

To Rotary Newmarket for their generous support at the outset of the project.

And of course to our wonderful families for their tireless support of our wild ideas:

Val & Paul Davis, Wendy Dickinson, Beth Davis, Amelia McBeth, Lars Filstrup Rasmussen, Elomida Visviki & Neneli Visviki-Rasmussen.

From all of us at Nanogirl Labs, we trust you will enjoy The Kitchen Science Cookbook, and hope that it brings you and your family many hours of exploration - and the unparalleled joy of learning and discovery.

Keep loving science!

Joe Davis
Executive Producer
Founder & CEO, Nanogirl Labs Ltd.

The MODELS

Once our recipes were written we set about working with Paul, our photographer, to bring them to life in front of the camera.

Thank you to each and every one of the families who contributed their time, effort and talent modelling for The Kitchen Science Cookbook. We are grateful to you all, both for enabling us to present this book full of wonderful images, and for the thoroughly enjoyable days shooting science experiments in the kitchen!

Appearing in The Kitchen Science Cookbook are:

Aya and Ali Al-Chalabi / Vivian, Annal, Dylan and Flynn Chandra / Val Davis and Amelia McBeth

Holden Gabriel / Kate & Zoë Hannah / Angel Jacobsen / Emily and Connie Lazarte-Simic

Kiri, Awatere, Te Aria and Kaiawa Nathan / Cindy Seaman and Amelia Lockley / Rudo Tagwireyi and Claire Shoko

Wendy, Amalie and Mieke Thompson / Christine & Hayden Wilson

The Kitchen Science Cookbook has been a worldwide collaboration. When we first started this project, we asked our friends on social media if they would be prepared to help us test our recipes. We honestly thought 5-10 of our personal friends might be willing to help. Thanks to the strength and generosity of our online community, our post was shared, and shared again - within 24 hours we had more than 2000 testers from all over the world who were willing to help us develop this book and bring science into people's lives.

To each and every member of our wonderful test community worldwide - thank you! We could not have done this without you. Your work has meant that we can say, with confidence, that every recipe in this book is easy to follow and uses only ingredients that can be found in most kitchens around the world.

Thank you!

Some of our wonderful testers...

Trent, Bethany and Finn Andrew / Ilaria Wright / Isabelle, Lewis and Hazel / Miranda Bull
Emma, Sam and Hannah Rich / Ellie, Katie and Sammie Shimsky / Joe and Spencer Sorge
Laura, Corey, Kyle and Abby Amerman / Owen, Mackey and JD Majewski
Aiden, Audrey and Eliza Stream / Finley and Adrienne Houck / Colart, Indigo and Sonja Miles
Zara, Elise and Simone Dury / Indira Bowden and Mum / Tamsin, Eleanor and Loretta Royson